D1460682

THE EXTRAORDINARY STORY OF LIFE ON EARTH

THE EXTRAORDINARY STORY OF LIFE ON EARTH

PIERO AND ALBERTO
ANGELA

Translated from the Italian
by Gabriele Tonne

Illustrations by Valter Fogato

59 John Glenn Drive
Amherst, NewYork 14228-2197

Published 1996 by Prometheus Books

00 99 98 97 96   5 4 3 2 1

Library of Congress Cataloging-in-Publication Data

Angela, Piero, 1928–
[Straordinaria storia della vita sulla terra. English]
The extraordinary story of life on earth / Piero and Alberto Angela ; translated from the Italian by Gabriele Tonne ; illustrations by Valter Fogato.
p. cm.
Includes bibliographical references.
ISBN 1–57392–043–6 (cloth : alk. paper)
1. Evolution (Biology) 2. Life—Origin. 3. Life (Biology) I. Angela, Alberto, 1962–   . II. Title.
QH367.A5413 1996
575—dc20 95–48159
CIP

Printed in the United States of America on acid-free paper

Contents

Introduction 7

1. From Night to Dawn
   January 1st (4 Billion Years Ago) 13
2. Taking a Dip among the First Living Creatures
   February 14th (3.5 Billion Years Ago) 43
3. Toward Complexity
   August 25th to September 10th
   (1.4–1.2 Billion Years Ago) 75
4. The Birth of Pluricellular Organisms
   October 27th (700 Million Years Ago) 99
5. In the Warm Seas of the Cambrian
   November 8th to 15th
   (570–505 Million Years Ago) 119
6. The Landing of Plants
   November 22nd (430 Million Years Ago) 147

7. The Landing of the Invertebrates
November 27th (380 Million Years Ago) 157

8. The Protagonists Arrive
November 27th (380 Million Years Ago) 167

9. Evolution on the March 197

10. The Earth Becomes Populated:
Amphibians and Reptiles
November 29th (355 Million Years Ago) 215

11. Before Dinosaurs
December 6th (270 Million Years Ago) 235

12. The Unknown Tale of the Therapsids
The Morning of December 7th
(258 Million Years Ago) 247

13. The Planet of the Dinosaurs
December 10th (225 Million Years Ago) 265

14. The Advent of Flight
December 18th (145 Million Years Ago) 283

15. The Rulers of the Cretaceous
December 19th (135 Million Years Ago) 295

16. An Explosion of Mammals
December 26th (63 Million Years Ago) 323

17. A Thinking Machine
11:50 P.M. December 31st (80,000 Years Ago) 355

18. The Next 4 Billion Years 377

Introduction

In one of his inspiring books, English biologist Sir Peter Medawar advised his young researchers to avoid bogging their readers down in mud or leading them barefoot across stretches of broken glass. He felt that a lot of suffering could be spared the readers of scientific articles or books simply by clearing the path for them.

This is certainly an intelligent (and considerate) way of looking at authors' relations to their readers, particularly those setting out on difficult ventures. And we have tried to respect this Golden Rule here because the route we have charted for ourselves and our readers is extremely complicated: it will cover the long and tortuous road that led to the emergence of life on Earth and to its evolution into ever more complex forms.

The format chosen—that of a logbook or journal—allows our two imaginary travelers to make firsthand observations and jot down notes about the events that took place on our planet in the course of four billion years of evolution. The voyage is simulated, of course, but our imaginations have helped us to overcome the restrictions of time, space, environments, and size.

The logbook condenses four billion years into a single year extending from January 1st to December 31st. This makes it easier to assess the relative distances between the various events and the duration of each era. For example, with January 1st established as the date on which life first appeared on Earth, it is now known that the only living forms existing until August were primitive unicellular organisms similar to present-day bacteria. Only in the second half of August did more complex cells with a nucleus appear. Life began to emerge from the oceans at the end of November. Dinosaurs appeared in early December. The mammalian era began in the last week of that month. *Homo sapiens sapiens* made his debut in the last minutes of the last day. Writing was developed in the last seconds.

During the first extremely long period, life gained momentum very slowly, like a language that first has to create its letters, syllables, and basic words before it can go on to build simple sentences, then more and more complex ones and finally burst forth in conversation, reasoning, and philosophy.

Our reconstruction of this long story is based on the work of researchers in the fields of biochemistry, paleontology, astrophysics, cell biology, zoology, and genetics, among others, who are intent on trying to piece together the various evolutionary events. It is also derived from the many fossil remains that have been found and that provide firsthand evidence of the creatures that lived during prehistoric times.

Finally, our reconstruction is based on observation of life on Earth today. Extant living forms represent an extraordinary kind of "time machine" for viewing the past: to use a somewhat unusual analogy, evolution is like a train from which passengers have continued to descend (or fall) for four billion years. Today we can still observe many of those passengers who got off at various stations along the way and who have remained similar to their ancestors: bacteria, protozoa, sponges, jellyfish, corals, worms, mollusks, fish, insects, amphibians, and reptiles.

Lined up, they form a huge retrospective show, or perhaps one should say a family album, providing an exceptional view—albeit through a somewhat distorted lens—of evolution. Then again, fossil remains have confirmed the similarities between modern and archaic forms. Although the exercise requires caution, it seems reasonable in some cases to try to deduce certain characteristics of past forms from observation of similar living forms today.

Several pieces to this grand mosaic of the evolution of life are, of course, still missing. Many areas, especially in the earlier part of the story, remain shrouded in darkness. But on the whole, superimposing and linking the results of the many areas of research investigating this subject from different angles provides a rather convincing overall picture, especially with regard to the order of events.

This volume ends with the appearance on Earth of the first hominids, our first direct ancestors (thus linking up with our previous book on human evolution*).

We would like to thank the following for their helpful suggestions during the writing of this book: Prof. Francesco Amaldi, biologist (University of Rome); Prof. Piero Cammarano (University of Rome); Prof. Steno Ferluga, astrophysicist (University of Trieste); Prof. Giacomo Giacobini, paleoanthropologist (University of Turin); Prof. Benedetto Sala, paleontologist (University of Ferrara); and Prof. Piergiorgio Strata, neurophysiologist (University of Turin).

A special thanks goes to Barbara Ferranti for the care and enthusiasm with which she prepared the various drafts of the book.

*Piero and Alberto Angela, *The Extraordinary Story of Human Origins* (Amherst, N.Y.: Prometheus Books, 1993).

NOTE

A word of caution. As will soon become clear, some parts of this book are more complex than others. The reason for this is that the emergence of life on Earth is not a simple story and, as such, cannot always be simply told.

We were, therefore, forced to make a choice of which we hope the more motivated and curious readers will approve: in some of the crucial parts of the story, we have gone into greater detail in an attempt to provide a more thorough explanation. We have preferred to run the risk of being a little difficult at times to giving in to the temptation of taking short cuts to please our audience, as they say.

For example, we felt it important to look more closely at certain fundamental aspects of the story and, in particular, to dedicate sufficient space to the first two chapters which deal with the origins of primitive life forms in the primordial sea. We did this not only because this period accounts for two-thirds of the entire evolutionary process, but also because it contains, in embryo, all successive stages of development. A good understanding of this stage is essential for an understanding of the basic mechanics of life. Should anyone feel bogged down, however, they may skip the more difficult parts and carry on—without detriment—to the rest of the story.

For our part, we have tried our best to put Medawar's advice into practice: hopefully the narrative device of the voyage and the use of our imaginations will keep our readers' paths free of broken glass.

Opposite page: The evolutionary time scale of life in the last half billion years. The preceding three and a half billion years (the Precambrian Period, which would take up a column seven pages long) simply encompass the evolution of bacteria and protozoa. After the end of the age of the dinosaurs, the names of the epochs are often used. The Paleogene is divided into the Paleocene, the Eocene, and the Oligocene; the Neogene is divided into the Miocene and the Pliocene; the Quaternary is divided into the Pleistocene and the Holocene, which extends to modern times.

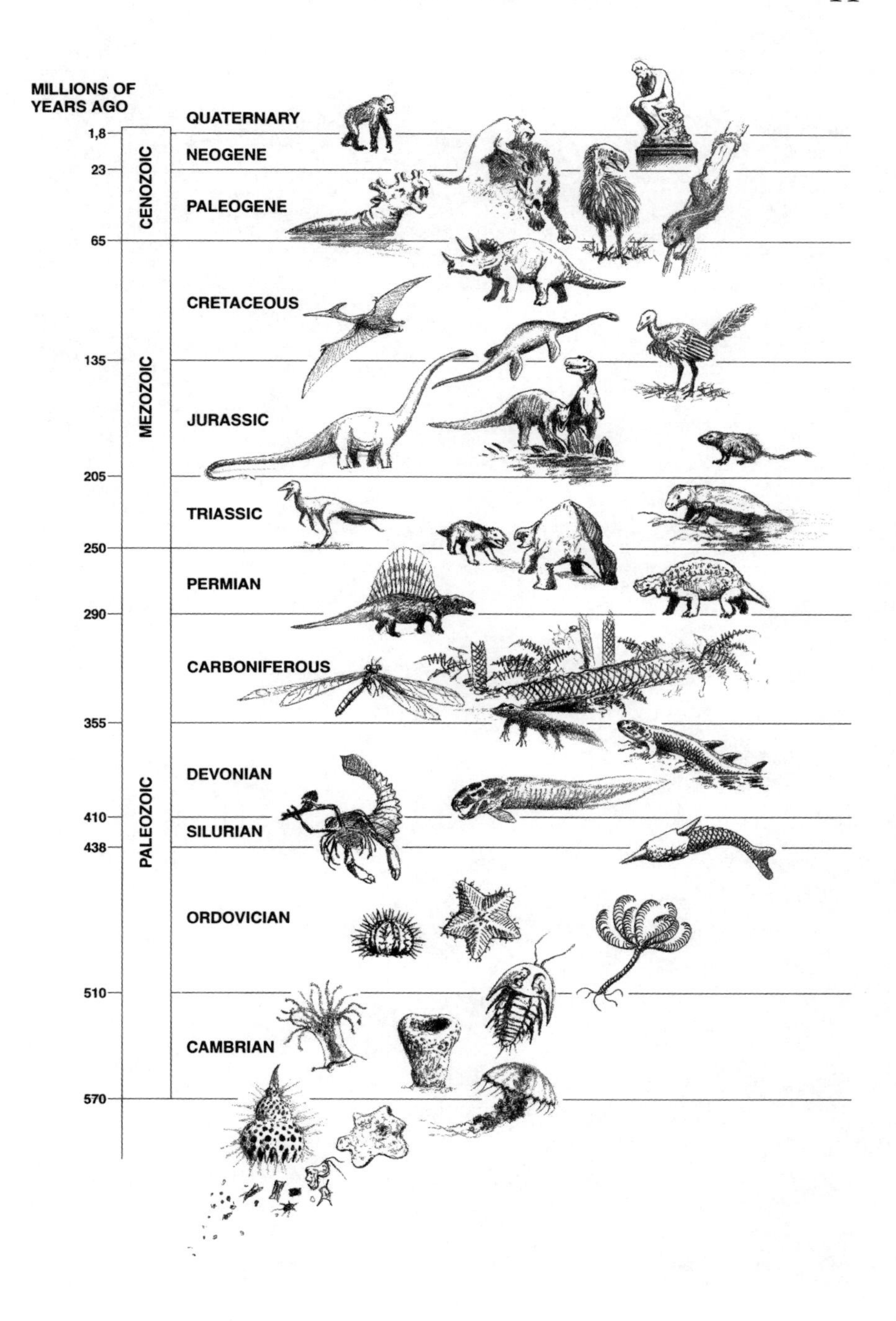
MILLIONS OF YEARS AGO
1,8
23
65
135
205
250
290
355
410
438
510
570
CENOZOIC
MEZOZOIC
PALEOZOIC
QUATERNARY
NEOGENE
PALEOGENE
CRETACEOUS
JURASSIC
TRIASSIC
PERMIAN
CARBONIFEROUS
DEVONIAN
SILURIAN
ORDOVICIAN
CAMBRIAN

1

From Night to Dawn

JANUARY 1ST (4 BILLION YEARS AGO)

As seen from an asteroid

Immense and silent, the earth gradually approaches. It looks like a pearl of strange hues, a jewel suspended in the darkness of the cosmos.

This earth is different from the one we are accustomed to seeing in pictures from space: its atmosphere is darkened by unusual shades of grey. At first sight, it seems to have a denser cloud covering, swept here and there into huge vortices typical of hurricanes.

The rocky fragment on which we are hurtling through space—an asteroid no more than twenty yards long—has just been "captured" by the earth's gravity, which is now pulling us into an orbit around the planet and gradually reeling us in toward the earth as if we were attached to an invisible fishing rod.

The asteroid will disintegrate when it penetrates the atmosphere, but that still leaves us a couple of orbits in which to observe the surface of the primordial planet.

The cloud cover beneath us is lit up intermittently from below. Immense storms flood the earth with rain, while spectacular flashes of lightning rend the skies. Even from up here, these storms look violent and the winds tempestuous.

From time to time, luminous traces streak across the clouds, fading into the darkness or bursting into cascades of sparks and streamers which in turn disappear. These meteors falling to the earth turn every night of this first day of January into the 4th of July.

From our position, the dense cloud covering the poles makes it difficult to tell whether they are ice-covered. This would not be impossible in theory, as some radar surveys show that even Mercury (which is so close to the sun that lead melts on its surface) has icecaps at its poles.

Glancing back at the sun, we are in for a surprise. Our star has rings around it like Saturn! Cosmic particles and bits of matter have been attracted by its gravity and shaped into an unexpected series of flat, concentric rings.

But that is not the only surprise. The sun is slightly bigger than it is today and is going through a period of intense explosive activity, emitting enormous quantities of radiation and particles in the form of solar winds. Not yet protected by its ozone shield, the earth is strongly affected by this outflow of radiation, which makes its surface an even more inhospitable place. But with time, the sun will shrink and take on its present appearance. The rings will also disappear, either swept away by the solar winds or drawn in by gravity like the scrap iron and wrecks of artificial satellites now orbiting around the earth. In fact, in this early period, the earth itself may have rings that we just can't see from our asteroid.

"Suddenly we find ourselves directly above one of the stars of this era: a huge, active volcano. . . ."

A lava desert

Finally the cloud cover opens enough in places to provide us with glimpses of the earth's surface. Vast lava deserts are split with apertures expelling fumes and vapors. Huge towers of solidified magma, i.e., molten matter beneath the earth's solid crust, rise amidst deep gorges and canyons.

These immense valleys have been formed by rivers swollen with the torrential rains that have pounded the earth for millions of years. Thick fog now conceals their lower depths from sight. If we could see Mars right now, it might have the same kind of valleys and rivers: in fact, space probes have revealed the existence of enormous canyons that were probably eroded by water.

Strange circular forms worn down by time stand out in the landscape below. These unusual amphitheaters are craters formed by the impact of meteorites or of now extinct volcanoes.

Almost everything we have seen on the earth's surface until now has been created by volcanic activity. Modern Earth has since donned a bright green coat of vegetation, but wherever that coat has worn thin or has disappeared, the desert-like landscapes reminiscent of earlier times reappear.

Suddenly we find ourselves directly above one of the protagonists of this era: a huge active volcano. The giant is miles high, a primordial Fujiyama. Rivers of incandescent lava flow from its central crater down its slopes, branching into myriad rivulets that spread out over the arid expanse at its base. One area the size of a city has become a huge incandescent lake of extremely hot, bubbling magma. It won't be long before it spills over into the valley below, submerging it.

We are flying over a very active part of the earth: one of those continental "zippers" that opens under the pressure of the magma, tearing open the earth's crust or separating huge portions of it.

This is the dominant mechanism on the earth in this period: rather like a thawing lake, in which huge sheets of ice collide. Here, too, huge slabs of the earth's crust, the size of continents, slowly move and push against each other (creating areas ravaged by volcanoes and earthquakes).

The colors are also different. Looking back over our shoulders, we see a huge sun hanging low over the horizon in a brilliant red sky. The hue is a much richer crimson than we see in our own sunsets: the particles suspended in the atmosphere amplify the sun's effect.

The light fades and our tiny asteroid carries us toward darkness. We are flying over the part of the earth now engulfed in night's shadows.

Although it is impossible to distinguish sea from land or cloud, we are graced with an exceptional view of erupting volcanoes, lava lakes and flows. Thousands of tiny lights and yellow flashes glitter on the landscape. It almost looks as though we are flying over immense cities. But the illusion is short-lived as a volcano starts to spew sparks into the sky.

The stars stand out vividly in the darkness of the cosmos above and around us. Given the lack of atmosphere, we can see an incredible number of celestial bodies, some of which are still being formed. Many stars have not yet come into existence.

We look vainly for two easily recognizable constellations, Orion and Ursa Major (the Big Dipper), and realize that many constellations have not yet taken shape or have a different shape from the one we know today. Orion and the Big Dipper will appear later; in fact, their stars are younger than our sun and even our planet.

This flight over the earth is an unforgettable experience; but unfortunately it will never be more than a flight of our imagination.

The show is not over, however, and as we gradually draw closer to the earth, our view of the incredible "fireworks" improves. In some places the night sky lights up in a web of stormy lightning. In others, the gloom is pierced by meteors which streak through the sky with long scintillating tails until they meet their end in a blinding flash of light. The spectacle is both terrifying and fascinating.

We are now rapidly losing elevation. Friction with the earth's atmosphere is already heating our asteroid and before long it will turn into a ball of fire. It's time for us to jump and pull the rip-cord of our parachutes.

JANUARY 5TH (3.8 BILLION YEARS AGO)

A primordial downpour

After a gentle landing on a vast and desolate plain of wind-levigated rock, we set out to explore our planet.

The air is probably a little warmer than in the modern world, while humidity is at the level of the tropics. Not a particularly hostile climate, on the whole, if it were not for the extreme differences between diurnal and nocturnal temperatures (as in the desert).

The smooth stones all around us are covered by a fine layer of sand and volcanic ash. This rocky landscape has been modeled by the wind which has accumulated deposits of tiny stones and has "sandpapered" larger reliefs. It strongly resembles our pebbly deserts, with its different colored islands (yellow, red, green, and brown, depending on the different mineral compounds) baking in the sun.

The most striking features are the silence and the immobility. It is as if time is standing still.

Suddenly we notice an almost imperceptible vibration underfoot, like the passing of a subway train: the ground is being shaken by the earthquakes that are a part of the earth's settling process.

Dark clouds start to scuttle across the sky. A cold wind has come up, suddenly lowering temperatures. We hear a dull rumble as the first drops of rain fall. Although we make a dash for it, a torrential downpour soaks us to the bone before we reach an overhanging rock forming a natural cave.

It would be nice to light a fire to dry our clothes and brighten up the place, but of course that is out of the question: not only is there no wood, but there is no oxygen in the atmosphere either. No fire can burn here.

The air is quite different from the air we know. The asteroids, meteors, and planetary fragments (stray rocks) that have fallen from space for hundreds of millions of years have brought various compounds (such as methane and ammonia, which later vaporized) with them in a frozen state. At the same time, the volcanoes have uninterruptedly spewed water vapor, carbon dioxide, nitrogen, hydrogen, and other gases (such as hydrogen sulphate, sulphur dioxide, and carbon monoxide) into the atmosphere ever since the earth's crust was formed 4.2 billion years ago. All these compounds have endowed the earth with a new atmosphere; the earlier one was swept away by the solar winds which removed the residual gases from the primordial nebula (composed mainly of hydrogen).

The deluge continues. If we think for a moment that this kind of precipitation has been going on regularly for millions of years, it is no wonder that the oceans we were flying over before were formed.

Where did all the water come from? Water, like the atmosphere and all other matter on earth, was brought in from space in frozen form aboard the stray rocks that accumulated to form our planet. The incandescent magma produced by the fusion of

"It is in this water, in this primordial ocean, that life originated."

those rocks released a large part of the water (up to 10 percent), but much of this was then converted to hydrogen after oxidizing methane. The residual water vapor was freed from magma together with other gases during volcanic eruptions.

This and the water freed directly into the atmosphere from the melting of the ice in asteroids and perhaps comets slowly created the lakes, seas, and oceans we saw during our flight.

Discovering water

The storm is over, but it has left the plain almost beyond recognition. Inundated with water, only the dark towers and lava masses—monuments sculpted and shaped by the wind—are still visible.

It takes us a few hours to cross the flooded terrain. Gradually a strange sound, like that of deep and regular breathing, becomes audible. Following it, we set off toward a row of low hills on the horizon and when we reach their summit, we are left breathless by the beauty of the boundless blue ocean before us, its waves lapping onto a black sandy beach. We scramble down over the rocks to bathe our feet in its water.

It is in this water, in this primordial ocean, that life originated. And this is the starting point of our adventure, our search for the first traces of that long process that led to the myriad forms of life, including human beings, that surround us today.

A promontory juts out at the far end of the beach. Beyond it, tongues of lava extend into the sea creating tiny fjords and lagoons. The interaction between rock and water has produced a kind of checkerboard of tiny protected pools and lakes irrigated by the sea. This seems to be the ideal place to observe what is going on in these waters.

As we approach one of the pools, we notice that the surface of the water has an unusual color and consistency. But we are disappointed when we kneel down to get a closer look: nothing is evident. There are no particles in suspension or movement of any kind. After a moment's hesitation, we dip our hands into the water. To our surprise, it's warm and has a slightly sticky or oily feeling to it.

Something is definitely going on here: something we cannot see but which could be exactly what we are looking for. We'll just have to change dimensions and scale down to molecular size to explore this world.

A trip among the molecules

We have scaled down to the size of a molecule, that is, a few millionths of a millimeter—a hundred thousand times smaller than

"Suddenly a blinding flash illuminates the scene and an explosion shakes the pond."

the thickness of a sheet of paper—and are now swimming around in this astonishing primordial environment. Entering into the water this size is rather like diving into a swimming pool full of styrofoam balls: the water molecules around us do not wet us, they simply pass us by like an immense moving crowd of people.

There goes a carbon dioxide molecule: an atom of carbon that carries two oxygen atoms around with it like a nanny. Carbon plays a very important role at this stage, even if there is not much of it around: in fact, carbon makes up only 0.034 percent of the earth. And yet, as all living forms are based on carbon chemistry, this element, which has its "arms" open, ready to catch other atoms and molecules as if in a merry waltz (or a continuous quadrille), is one of the leading actors in the story of the construction of life on Earth. In fact, carbon, along with hydrogen, oxygen, and nitrogen, is the central element of all life, from the bacterium to the rhinoceros, from the frog to the eagle, from a blade of grass to the human being.

In the midst of the molecules passing by, we easily recognize the "organic" molecules that have rained down from the skies and are composed of various combinations of the four elements. Their numbers are increasing rapidly as a result of the continuous chemical reactions taking place in the primitive atmosphere.

As mentioned earlier, the extremely luminous atmosphere above us is composed of a very different mixture of gases from that which envelops the modern earth. This primordial mixture would have killed us immediately. But these poisonous gases provide an inexhaustible source of raw materials for the formation of complex molecules triggered by the various forms of energy constantly wracking the environment: lightning and electric storms, radioactivity, volcanic eruptions, meteorites, and, above all, ultraviolet (UV) radiation. There is no ozone in the sky, and the lethal UV rays rain down like bullets.

These reactions are well known and have been reproduced in

the laboratory. Stanley Miller's* famous experiment has proven that if the gases constituting the earth's primitive atmosphere (hydrogen, methane, ammonia, water vapor, and so on) are placed together in a closed vessel and the energy of lightning is simulated by electrical discharges, a large number of organic compounds—all those that can be formed (as in a game of Lego blocks) with the four elements present (carbon, oxygen, hydrogen, and nitrogen)—will soon form.

Naturally, the most abundant compounds are those made up of simple molecules. The least abundant are those composed of more complex molecules (that is, with a greater number of atoms). But even some amino acids are formed (after a series of reactions), and it is well known that amino acids are the "building blocks" for proteins; that is, they are the basic materials needed to form muscles, skin, nerves, and all our organs, including the brain.

The amazing thing is that there are only twenty different kinds of amino acids (fewer than the letters of the alphabet) in living organisms. But by bonding in various ways, they can generate a potentially infinite number of combinations, giving rise to the enormous variety of cell functions and structures found in the living world.

Now we can distinguish the amino acids more distinctly as they float around in this sea of molecules: all are made up of various combinations of the four basic elements: carbon, oxygen, hydrogen and nitrogen; only some—in fact, only two of the twenty—also contain a sulphur atom. They all exhibit the same basic structure: nine atoms plus a lateral chain whose length and shape vary in different amino acids.

Actually, there seem to be far more than twenty amino acids drifting around here, but the other "formulas" are not destined to take part in the biological Lego game. Only the most suitable will be used and will continue to appear in life's assembly kit.

---

*Stanley Miller, biochemist and author of pioneering studies in the 1950s.

Lightning bolts

Suddenly a blinding flash illuminates the scene and an explosion shakes the pond. We are thrown violently as a wave of intense heat surges over us. Lightning has struck the pond! We are sucked violently into a vortex, only to be spit out again at the surface of the pool.

From here, we can see the molecules in the atmosphere: some large ones are gradually sinking into the water and drifting to the bottom. One strange molecule in particular attracts our attention as it is sucked into a whirlpool. It comprises the four basic elements—carbon, oxygen, hydrogen, and nitrogen—but is assembled in a completely different way. It is an *adenine molecule.*

The interesting fact is that adenine is one of the four basic components (the others are thymine, guanine, and cytosine) of DNA, the double helical molecule that carries the genetic heritage of all living creatures. This means that we are witnessing the birth of a piece of DNA.

But where has the adenine come from? Laboratory experiments simulating the conditions of the primitive atmosphere have shown that large amounts of hydrocyanic (prussic) acid are formed by electrical discharges. This particular acid spontaneously "polymerizes" in aqueous solution; that means, it joins up with other molecules, giving rise to a more complex molecule. And that complex molecule is adenine (precisely a pentamer of hydrocyanic acid).

The fact that adenine can be formed spontaneously strongly supports the hypothesis that the first molecules of life were assembled in the primordial sea.

All this organic material, originating in the atmosphere and jolted by lightning and other forms of energy, has been accumulating in the oceans and above all in the primordial lagoons, where it is generating a kind of primordial or "prebiotic" soup in which billions of molecules have reacted continuously for millions of years.

But is this really enough to explain the progressive assembly that led to the spontaneous appearance of life on Earth? To the appearance of cells or "spherules" (albeit very simple ones) able to build themselves using the raw materials in their surroundings and able to remain stable in a changing environment (thanks to a primitive metabolism)?

The answer is not that simple. Researchers like Aleksandr Oparin, J. B. S. Haldane, Harold Urey, Stanley Miller, Leslie E. Orgel, S. W. Fox, Ciryl Ponamperuna, and many others have made fundamental contributions to the study of the prehistory of life. They have devised fascinating experiments and formulated hypotheses, but the mysteries surrounding these earliest stages of evolution have still not been cleared up.

This is the most difficult stage of the whole story. It is also the period about which we know the least. The following pages (up to the end of this chapter) are, therefore, those most likely to be strewn with difficulties. But they are also the ones that tell one of the most fascinating parts of the story, and we therefore think that it may be worth a try to read them.

Traveling at millions of miles per hour

A gust of wind suddenly sweeps us off our feet and, together with billions of molecules, carries us along at speeds of up to 25–30 miles per hour (that is, 35–45 feet per second). From our perspective, however, it seems as though we are traveling at millions of miles per hour: given our size (that of a molecule) we are, in fact, covering a distance equal to 150 million times our height every second.

Swept along with us in this dizzy flight are what look like huge boulders, actually grains of sands. Terror grips us as we see a huge mountain loom up ahead. Only seconds later, the boulders below us smash into it with phenomenal violence, breaking off chips of rock (this is how wind erosion occurs in nature). Our hearts stand still as we careen like kamikaze pilots; but then, just in time, the wind buoys us up and past the mountaintop. Only afterward do we realize that it was no more than a large rock on the beach.

As the wind lets up and the gust starts to lose force, we find ourselves in a sheltered area, perhaps an inlet nestled between two rocks. In the midst of other suspended material, we swirl around in large, silent vortices as the "boulders" begin to fall to the ground.

An enormous number of organic molecules are mixed with the atmospheric gases. Most of them are *monomers* (that is, relatively simple molecules), waiting to be transformed into more complex molecules (called *polymers*) like freight cars waiting to be hooked up to form a train.

We settle gently to the ground in a hilly area strewn with huge boulders—actually grains of sand on an isolated beach. Other molecules drift around. Bewildered, we start to explore this fantastic microcosm, unable to grasp the meaning of what we are seeing. The only thing that we can conclude is that the immense "Lego game" being played at the infinitesimal level in the atmosphere first and later in the oceans is still undergoing constant change. Various types of agents—in particular, lightning, ultraviolet light, different kinds of radiation, shock waves, and hot volcanic ash—are causing molecules to unite, break apart, fragment, and become more complex.

Climbing onto one of the higher boulders to get a better view of the lagoon, we wonder about the probability of this "Lego game" giving rise to life.

Assembly and its adversaries

The first step toward life seems to be the transition from a simple molecule to a complex one, that is, a polymer (such as DNA or proteins). In fact, the mere presence of the basic components of such molecules (amino acids for proteins, nitrogen bases and other components for DNA) in the prebiotic soup is not enough. They must unite. Otherwise it would be like having thousands of pieces of Lego but no way of putting them together.

This step toward polymerization is an extremely difficult one: years of research and experimentation have shown that it is hindered by all kinds of "adversaries."

1. The first "adversary" keeping the freight cars from hooking up into a train (polymerization) is the lack of sufficient *free energy,* that is, usable energy: some steps of the chemical assembly (such as the production of nucleotides) would not take place spontaneously in the prebiotic soup because they require additional energy.

2. Another, more traditional "adversary" of the construction of life is *ultraviolet light.* In the absence of a protective shield of ozone, ultraviolet rays are destructive to all kinds of life. Only underwater, at a depth of at least thirty feet, do UV rays lose their "lethal" action.

3. A third and equally traditional "adversary" is *oxygen.* In the dynamics of chemical reactions, oxygen molecules play a predominant role (especially at this stage). When they come on the scene, they act like certain male dancers who snap up certain women as soon as they set foot on the dance floor (or, if you prefer, like policemen who handcuff and whisk away criminals). More specifically, free oxygen bonds easily with carbon (which is

one of the basic elements of life) and inhibits the formation of more complex molecules by causing oxidation.

In modern biology, the presence of oxygen in the atmosphere is no longer a problem (on the contrary, it is an advantage); but in this early period, free oxygen snapped up the molecules and was, therefore, a very dangerous element that had to be avoided—or neutralized.

Fortunately, there was very little oxygen in the primordial world. Had it been otherwise, no "train" could ever have formed by progressively hooking up more cars.

Organic molecules from space?

The difficulties just described (especially the first two) make it seem unlikely that the spontaneous assembly of life could have taken place directly through accidental bonding in the prebiotic soup. Years of research suggest that the road taken was a little more roundabout and involved some kind of trick, which is still unknown.

Completely different approaches have also been attempted, hypothesizing, for example, that biochemical assembly took place under particular environmental conditions such as freezing temperatures or drought, or in certain clayey soils which could, in theory, facilitate the formation of amino acid chains.

To make the event of spontaneous assembly more probable, some have even thought of extending the "Lego game" to the entire universe, suggesting that life, or rather, certain complex molecules, may have arrived directly from space. It is, in fact, known that many organic molecules exist in space: they have been detected in interstellar space and have been found in abundance in certain large carbon-based meteorites that have fallen to Earth (not only the amino acids found on Earth, but also some that are not: as many as thirty-five of the latter were identified on the so-called Murchison meteorite).

All this seems to provide evidence that the basic chemical building blocks of life can be formed relatively easily in the universe. Why, then, could rather complex polymers originating elsewhere in the universe not have been brought to Earth by rocks or asteroids? Such an event would have increased considerably the probability of the right biochemical assembly.

The idea is not very popular among the scientific community (and even less so, needless to say, is that of a "cosmic panspermia" directed by intelligent extraterrestrial beings). One of the reasons for this is that it would do nothing but shift the problem of the origin of life to other regions of the universe.

Is there any way, then, to understand how life began? If it does not seem likely that molecules could assemble in the atmosphere or in water, if the *specific energy* for certain reactions was lacking, if numerous obstacles hindered these first steps (especially polymerization), what biochemical route did the first primordial cells and life take?

Our thoughts are turned suddenly from these problems by a deafening noise which seems to be increasing in intensity. We barely have the time to look up to see a wall of water barreling toward us. From our microscopic point of view, the wave looks hundreds of miles high: a veritable mountain of water, immense and terrifying.

It crashes down upon us, wrenching us away along with the soil, the "boulders," and the molecules around us. Razing everything in its path, it tumbles and rolls us into a violent current of millions of water molecules.

Actually, this is no more than the tide encroaching on the shore of the lagoon where we have landed. A strong undercurrent first sucks us toward the bottom and then pushes us forward, like

---

Opposite page: Diagram of how phospholipids can combine to form a membrane which, under the action of wave movement, can produce a sphere.

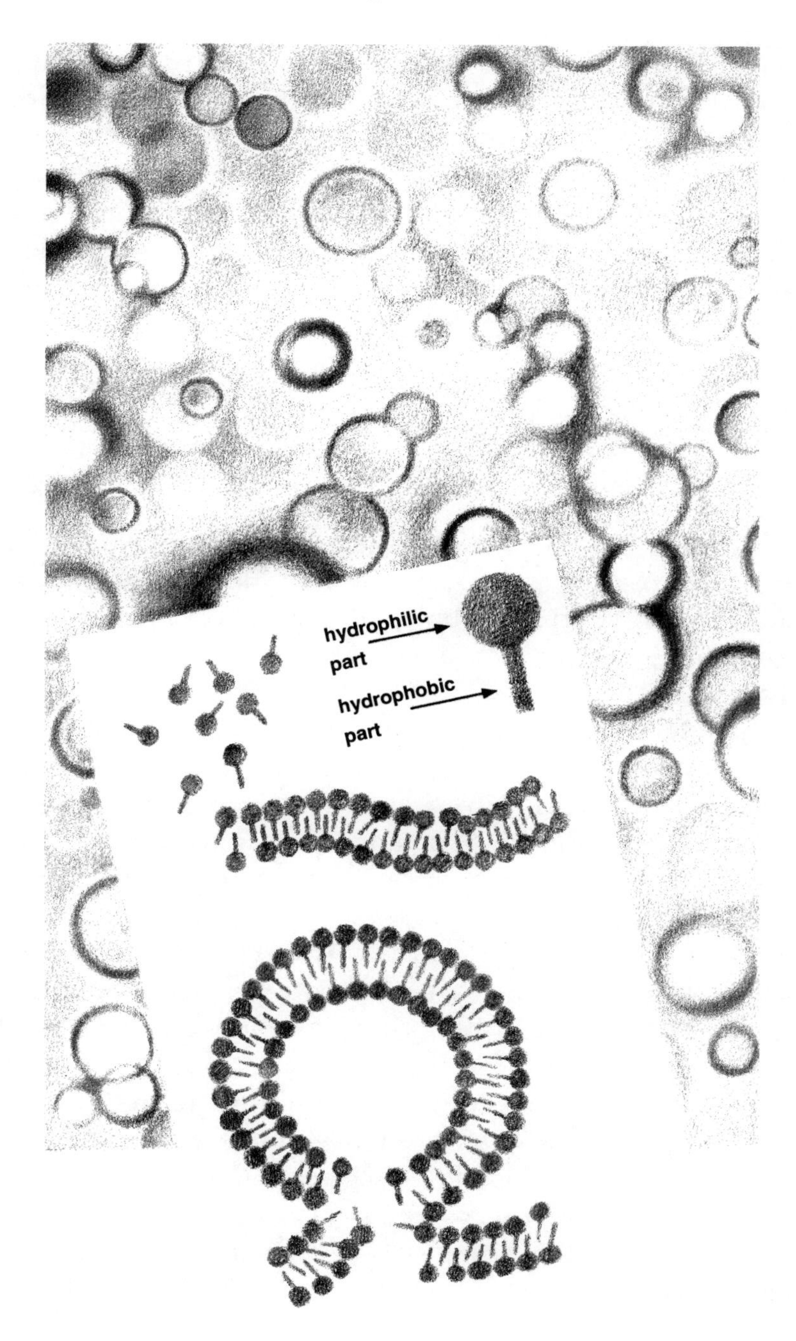
hydrophilic
part
hydrophobic
part

a huge churn. Molecules race past us in all directions. Most are quite different from the ones that surrounded us previously. Driven by the wind and the waves, many organic compounds have concentrated here in these coastal fjords and lagoons.

Suddenly, something totally new and bizarre, something we have never seen before, takes place: not far from us, under the pressure of the waves, chains of molecules unite to form a sphere. Why has this happened?

The birth of a membrane

If you think about it, the phenomenon is not totally foreign to us: it takes place every time we pour a little oil into some water and shake it. We create bubbles. Today, as in the past, little balls of oil form spontaneously by means of mechanisms well known to all.

The drawing on page 31 illustrates what we just saw. Some molecules have two parts: a part that attracts water (*hydrophilic*) and a part that repels it (*hydrophobic*). When a large number of these molecules come into contact in an aqueous environment, they tend to unite in a special way: the hydrophobic parts draw close together and the hydrophilic parts reach out toward the water. In this way, they form two-layer chains which have the natural propensity, when shaken by wave movement or whatever, to turn in upon themselves, forming a membrane that closes like a sphere.

Such molecules are very common in nature today: they are the phospholipids (of fats), which make up more than 50 percent of the material contained in our cell membranes.

But where did this fat come from in the primordial sea? Once again, it was probably brought to Earth on rocks drifting around space (asteroids and tiny planets) which contained large amounts of frozen methane. When the methane thawed into the ocean, it was attacked by ultraviolet rays and transformed into a huge layer of condensed hydrocarbons. Researcher Dick Holland of Harvard University believes that the layer on the primordial oceans may have been several yards thick. These hydrocarbons were then slowly demolished by solar radiation, partially oxidized, and rendered soluble.

At the end of this process, large oily patches were left on the oceans' surfaces (this explains the sticky sensation we experienced when we first dipped our fingers into the lagoon). As a result of the action of the waves and rain, these oil slicks finally gave rise to an infinite number of spherules or vesicles.

We cautiously approach the spherule or, rather, lipid sac, which is bouncing around gently in the water. Even though it is no more than a few thousands of a millimeter in size, it looks immense to us.

Could this be the fundamental "invention" behind life? Indeed, this rather simple and spontaneous process creates places that could favor the advent of life; a cavity of this size can host a large number of molecules and provide a place for them to interact, like a test tube.

In fact, chemists use test tubes to bring into contact the various components that they want to react. Without a container of this kind, molecular encounters and reactions would be diluted in the environment and, therefore, lost.

Exploring the great sphere

Like scuba divers exploring the submerged remains of a transatlantic liner, we swim up to the huge lipid vesicle. We are at a depth of approximately thirty feet, close to the sea floor. At first, we swim slowly around the vesicle to explore its surface.

What does it contain? Who knows? It looks impenetrable. (These lipid vesicles do not yet have "pores" which will be formed later by proteins adhering to the membrane). At the moment, only small lipid-soluble molecules like carbon dioxide can get through the membrane. But the secret of life may well be concealed inside it.

Attempts have been made in the past to simulate chemical reactions inside oily bubbles prepared in the lab. By introducing an enzyme (phosphorylase) into an artificial oil drop and allowing it to react with glucose molecules passing through the membrane, a starch was obtained which made the sphere swell (and, at some point, "divide"). As we all know, growing and dividing are two basic qualities of life.

These situations are artificial, however, and the main criticism of them is that enzymes (which are proteins and, therefore, complex molecules) did not exist at the time of the prebiotic soup.

Could this action have been carried out by some special catalysts or "coupling" molecules? Or some other kind of molecule? How could there have been an exchange between the outside and the inside of the vesicle if it had no pores (as there were no proteins in the initial phase)?

These basic questions are indicative of the difficulties modern scientists come up against in trying to understand what went on inside those lipid vesicles. Even the assembly of simple molecules (polymerization) is a problem, as the vesicle contains water: if the water is neutral or basic, certain kinds of assembly cannot take place because the molecules would repel each other. Reactions of this kind can only occur in a strongly acidic environment. Once an internal acidic environment has been created, polymerization (and phosphorylation) will take place spontaneously.

But how could the liquid inside the vesicle have become acidic? A source of energy—usable energy—is required. The problem is finding out where that energy could have come from.

All these matters give a further idea of the complexity of the problems and the difficulties that researchers face in trying to reconstruct such an ancient event in the totally different environmental conditions that exist today. There is also the question of whether these events could have taken place at all. What is the probability of such spontaneous assembly, not only in the initial stages, but at all stages? For example, what is the probability of the nucleotide sequences forming the first genes?

Unfortunately, the number of sequences that could theoretically be formed through assembly is so high (indeed astronomical) that no probabilistic calculations can be carried out. It would be like trying to win a football lottery with over one million matches on the coupon.

Can no way be found out of this dilemma?

The secret phrase game

The problem may have to be viewed from a different perspective. The famous paradox of the monkey and the *Divine Comedy* comes to mind: how long would it take a monkey to write the *Divine Comedy* by hitting the keys of a typewriter randomly? Although theoretically possible, it would apparently take such a long time that the probability of the event ever occurring is practically nil.

But the problem can also be seen from a totally different point of view. C. E. Shannon, one of the founding fathers of the information theory, subjected a group of his students to an unusual test. They were supposed to guess a secret phrase by suggesting letters to which the examiner could reply only "yes" or "no." For example, if the phrase was "The cat is walking on the roof," they were supposed to guess, through successive trials, the twenty-four letters in the right sequence. If it had been tackled in a completely random way (that is, drawing the letters of the alphabet from a bag), the game could probably have lasted forever, given the almost infinite number of possible combinations.

But language has its own logic: there is a balance between vowels and consonants, between syllables and the association of syllables. In other words, anyone wanting to guess the phrase would never try to put twenty-four "a"s in a row, or twenty-four "b"s or "c"s. In this specific case, after having guessed the initial "t," most people would try an "h." Having guessed the "c" and the "a," most people would immediately guess "cat" (or "can," etc.), substantially reducing the number of guesses required for solution. All crossword puzzle fans know these shortcuts.

Chemistry has its logic, too. Associations are not purely random: there's a "grammar" to chemistry as well. In fact, when Stanley Miller had electric discharges pass through test tubes full of gases similar to those of the primordial world, he obtained organic molecules that obeyed these grammatical rules. In a word, there is a natural tendency for molecules to aggregate in certain ways—and not in others—under certain circumstances.

This means that the evolution of the primordial molecules was much less a matter of pure chance than may initially seem the case. It was a response to environmental conditions and the dictates of the laws of chemistry. This is especially true if one considers that conditions of "selection" have been present from the very beginning of the construction of life on Earth, even among molecules (they have now been verified in the lab).

In this regard, the German Nobel Prize-winning biochemist Manfred Eigen has invented an ingenious "evolution machine" in his laboratory at the Max Planck Institute. Made up of physics and chemistry components, it accelerates the "evolution" of a virus, allowing for simulation and study of the successive mutations of DNA sequences.

Darwin proposed that the more complex was generated by the less complex by means of natural selection. Eigen writes, "Why should this principle not hold true for the complexity of macromolecules?" Here, too, competition (and "cooperation") have eliminated an enormous number of accidental mutations and improved and stabilized those selected.

While Eigen thinks that it is impossible for us to reconstruct the various steps in the evolutionary process, he does believe that we can understand the natural laws that must have governed those steps.

Meantime, back in the primordial lagoons, the lipid vesicles—protected from ultraviolet light by thirty feet of water and constantly replenished with new organic material washed up by the ocean—are gradually evolving into increasingly efficient and lasting spherules. They were not yet cells, but they certainly resemble them. They constitute a kind of "missing link," if you will, between nonlife and life.

Production of themselves

It is evening before we return to our little inlet, leaving the world of molecules and lipid vesicles behind. The sun has slipped behind the horizon and the stars are glittering in the distance.

We will be returning to these seas before long to see how evolution is coming along. But now, as we gaze at the stars, our thoughts wander back to what we witnessed, especially during the most recent part of our journey. Both of us were impressed by the formation of the lipid vesicles. We felt as if we were witnessing a fundamental step in the history of life: the birth of an "individual."

In fact, their membranes make these vesicles autonomous entities, capable of having exchanges with the environment and of reacting to it. The vesicles are separate systems able to establish boundaries between themselves and others.

All of this is extremely elementary and does not substantiate use of the term "living system": what we saw was no more than a vesicle full of organic molecules, some of which are very reactive. But we can almost see the increasing ability to construct components that will respond in a more and more complex manner to the environment and will ultimately become highly efficient enzymes capable of accelerating reactions a million- or billionfold.

This is the beginning of metabolism (or to use the technical term, *autopoiesis,* meaning the "production of oneself"). Metabolism will allow living structures to remain stable in a changing environment and to counter, if only temporarily, the tendency toward deterioration and disintegration (in other words, death) dictated by the Second Law of Thermodynamics: the tendency toward greater disorder in organized structures known as "entropy."

The living and the nonliving: the first light of dawn

It is difficult to establish when the transition from chemistry to biology, or rather, from the nonliving to the living took place. It all depends on how you define life. It's rather like the transition from night to day. At what precise moment can you say that it is day? Someone has suggested that it is when you can read a newspaper. Fine, but does that mean when you can read the headlines or make out the fine print?

The problem is similar for the evolution of life. It is easy to distinguish between midnight and midday, but dawn is an intermediate stage that is difficult to graduate. The transition from chemistry to biology (that is, from the nonliving to the living) passes through the stage of biochemistry. What we observed during our dips into the prebiotic sea was the progressive preparation of the "assembly kit" destined to construct slowly forms that will, at some point, be called life.

But our knowledge is still in its initial stages. We can barely make out some of the headlines of the story of life, no more. We are still far from deciphering the articles.

How long will it take us to understand what happened in the lagoons four billion years ago? Scholars studying the problem feel that the conditions on the primitive earth were suitable for life to arise spontaneously: all the necessary basic molecules could form naturally (or by chain reaction) thanks to the presence of chemical compounds and energy in the environment. The same scholars also agree on the fundamental role played by the lipid vesicles or spherules, which created a protected setting for the evolution of primordial biochemistry.

Where hypotheses diverge is on the details, that is, the sequence of events. And this is because the process cannot (yet) be reproduced in the lab.

In other words, many of the components have been identified and many of the assembly sequences seem very convincing; however, the puzzle is still far from being completed. It is as if we were given a thousand pieces of a kit and asked to construct an unknown machine without an instruction booklet. We might even succeed, but it would certainly take a long time.

The problem of solving this enigma is complicated by the fact that few people work in the field: with no industrial, military, or pharmaceutical spinoffs, this kind of research receives limited resources. The problem is common to all sciences exploring the past, including paleontology and palaeoanthropology. There is much curiosity, since these studies attempt to shed light on our origins and on the great venture that led to the appearance of the human species on Earth, but financing is still scanty.

The interesting thing is that it would not take much money: nothing compared to high-energy physics, which requires cyclotrons, or even to the "mapping" of the human genome. Costs are extremely low and a minimum of investments would be enough to activate skills and minds and obtain more important results. It would also attract young people to this fascinating field of research.

We decided to put this little "message" into a bottle in the hope that some sailor may pluck it from the ocean.

Under the primordial moon

The air has become cold and damp. It is raining far out over the sea and we can see lightning bolts dropping out of the sky on the horizon. The clouds are moving in to block out the stars, but the moon is still visible. It looks strange, however: it is amazingly large and void of its usual markings.

In fact, in this primordial period, the moon is much closer to the earth. It looks like an enormous sphere hanging over our heads. It is also moving at much greater speed. That is why the tides are so strong, covering and uncovering immense stretches of beach and coast land. And this is how the coastal pools and lagoons are replenished with water and new organic material.

But the most extraordinary consequence of the moon's proximity is its effect on the rotation of the earth: the tides, caused by its gravitational pull, act like a "brake" on the earth, gradually slowing down its rotation over the course of millions of years.

As a result, days and nights are much shorter in this early phase. With the passing of time, the moon will gradually move away from the earth, but it will continue to act as a gravitational "brake," causing days and nights on Earth to become increasingly longer.

Astronomers predict that this braking action will continue in the future (even though it will become weaker and weaker) and that days on Earth will continue to lengthen. Until they coincide with lunar days, that is, twenty-eight days on Earth at present? Although calculations are complex, that seems to be the trend, but it will take a very long time.

Our day ends with a last look at the heavens—so different from the sky we are accustomed to—with its immense moon, unfamiliar stars, atmosphere lacking oxygen, and more intense sun raining lethal ultraviolet radiations down on an earth still unprotected by an ozone shield. For many millions of years, this desolate and savage landscape will characterize our planet. Yet, the atoms that will compose plants, animals, computers, automobiles, and human beings are all already present on this primordial earth, scattered throughout the oceans, the volcanoes, the rocks, and the atmosphere. They are waiting for a slow chemical, biochemical, and finally cultural process to assemble them—as in a gigantic Lego game—into increasingly complex forms: a process of constant construction and destruction of beings and things, from bacteria to sequoias, from jellyfish to humans.

This is the part of our story that we will be exploring in the next chapter, beginning with the ancestors of bacteria.

“But what it illuminates comes as a total surprise: this liquid stage is anything but deserted.”

2

Taking a Dip among the First Living Creatures

FEBRUARY 14TH (3.5 BILLION YEARS AGO)

The first encounter

Approximately 500 million years have passed and we have come back to take another dip in these lagoons connected to the open sea to see exactly what is going on.

Nothing seems to have changed in this liquid desert. The landscape is as lifeless as ever. These surroundings, so different from the ones we know on "our" earth, make us feel a bit uneasy: there are no fish, no shells or coral, not even tiny jellyfish or wisps of algae. All we can see is an endless expanse of sand and the dark profiles of volcanic rock.

A shaft of sunlight cuts obliquely into the waters of the lagoon, like a spotlight onto an empty stage. But what it illuminates comes as a total surprise: this stage is anything but deserted. It is teeming with actors. Life is pulsing around us—life that is not yet visible to the naked eye, but that is already active and varied.

Who are these new protagonists? They are the most minuscule and ancient creatures to appear on our planet: the *prokaryotes* or *prokaryotic cells.* These primitive cells are little more than fluctuating vesicles containing primordial filaments of DNA, but they are already capable of performing some fundamental vital functions, including reproduction.

To unveil some of their secrets, we are going to have to explore them from the inside. So we scale down to microscopic size once again and set off toward them.

At approximately thirty feet of depth (the depth at which life is protected from ultraviolet rays), we are nearing the sea bottom. Just below us, we can make out an immense mass of moving microorganisms. Each is extremely small, only a few thousandths of a millimeter—hardly any bigger than the spherules we encountered in these same waters many millions of years ago.

As we draw closer, these living forms appear in all their splendid simplicity and efficiency (it is now "broad daylight" on the evolutionary scale). Coming alongside one, we reach out to touch its outer membrane, which is firm and resistant and pockmarked with "windows"—pores—providing entry.

Taking advantage of one of these passages through the membrane, we enter the cell and find ourselves in a dense liquid, the cytoplasm. This is already very much like the cytoplasm found in our cells today, but everything else is different, much less elaborate. It would be like comparing a barren room with only a little furniture to a lavishly decorated apartment. The few pieces of furniture that are present, however, are the ones needed for survival.

Continuing toward the center, we come across macromolecules, enzymes, and RNA, but none of the organelles found in more evolved cells. There isn't even a nucleus to enshrine the genetic information: the DNA filament is simply condensed in a region of the cytoplasm. This is why these cells are called *prokaryote*: *pro* means before; *carion* means nucleus.

A primeval battery

Suddenly a real gem appears: a molecule that constantly assembles and disassembles itself, ATP. It is a kind of rechargeable battery that supplies the cell with the energy needed for its vital processes. This tiny chemical marvel is so perfect that it is still present today in *all* cells of *all* living organisms.

ATP (the abbreviation for *adenosinetriphosphate*) supplies energy for an infinite variety of cell activities: from the most primitive, such as the transport of molecules across the membrane of a bacterium, to the most sophisticated, such as those required for the movements of a ballerina. Its secret lies in its ability to "load" a phosphate group (thanks to an enzyme that breaks down glucose, that is, sugar) and then "shoot" it upon request, thereby producing energy.

Where do these simple but extremely efficient cells come from? How have they evolved up to this point? No one is in a position to answer this question today. These cells are, in fact, separated from the first organic molecules we saw in the primordial oceans by hundreds of millions of years of darkness: a long evolutionary course that has still not been clarified and whose many stages will probably not be completely reconstructed for some time to come because of the lack of fossil evidence.

Of course, it is not unreasonable to think that DNA and other kinds of molecules may accidentally have been contained inside the membranes which were forming spontaneously in the primordial seas (as described in the previous chapter). And that fortunate evolutionary occurrence led to the progressive transformation of these vesicles into protocells and, later, the prokaryotic cells before us.

Top: A fossil bacteria (prokaryote cell) called *Eobacterium isolatum*. It comes, as does the filament of organic matter in the lower righthand illustration, from a Precambrian deposit in Fig Tree, South Africa. It is approximately one half of a thousandth of a millimeter long and dates back 3.2 billion years. The group of bacteria shown lower left is younger: it is only 2 billion years old and comes from the Gunflint deposit located on the shores of Lake Ontario.

Tiny duplicating planets

We continue our exploration of these shallow waters, immersed in a myriad of suspended microscopic globules that drift on the currents like tiny planets in space.

One is of a particularly strange shape; more oblong than round, it has a funny kind of narrowing in the middle that makes it look rather like a peanut. This is a prokaryotic cell undergoing division.

Let's focus on the process a moment. This is one of the fundamental moments in the history of life: replication, that is, an organism's ability to reproduce itself.

The mechanism is still very simple here: the cell simply duplicates by producing an identical copy of itself. Inside, the DNA filament doubles and divides, concentrating in the cell's two extremities. The "waist" of the cell slowly narrows, as if squeezed by an invisible hand, until two separate entities come into existence.

The prokaryotic cell is undergoing the final stages of division. Its "waist" is extremely narrow; the cell now resembles an hourglass. Slowly the membranes close off and the two parts separate, creating two identical individuals—twins.

This is the way in which millions of billions of primordial organisms (ancient bacteria) started to multiply and populate the waters of our planet, adapting and differentiating in the course of eons. But how did they acquire this ability?

Of course, we do not know exactly what kind of transformations these first cells underwent, but we do have some indirect clues. The requirements of creatures living in water today are probably not very different from those of yesterday, particularly in the matters of food, motion, shelter, and reproduction. Therefore, it is probable that by observing the extant prokaryotic cells (Bacteria, Archaea), we can get an idea of what life was like in the primordial oceans.

On Earth today, prokaryotes are found by the billions in all ecological niches: in mud, the atmosphere, hot springs, glaciers, underground, and, naturally, water. They still lack a nucleus, and possess only a very simple cytoplasm surrounded by a primitive membrane. And they are still the same size as the primitive prokaryotes: a few thousandths of a millimeter. With due caution, observation of their contemporary descendants may be used to infer certain features of primordial prokaryotes.

Primordial systems of propulsion

It is very likely, for example, that ancient prokaryotes started to develop "oars" to help them move around more easily in the water. Bacteria today display a wide variety of such solutions. Although some have no means of locomotion—they are simply tiny drifting planets—other have flagella, that is microscopic tails, which function like small outboard motors. Flagella can be either single or grouped, highly developed or rudimentary; or they can be made up of a series of more organized cilia, i.e., short hairlike projections. The cholera vibrio, for example, has from one to three cilia at its extremities; the tetanus bacillus has fifty to one hundred lateral flagella.

It is enough to examine a drop of water under the microscope to see the ease with which certain microorganisms move in water. Increasing magnification to look at a single flagellum reveals that it is made up of simple protein molecules (fibrils) that hook together. No muscles, no nerves, no tendons. Unlike legs or fins, these flagella are extremely simple structures, and yet the slipping and release mechanisms of the molecular chains of which they are composed make them the first "limbs" capable of effectively executing a swimming stroke.

Were the first prokaryotic cells that populated the earth billions of years ago similar to these? We do not know, even though it seems reasonable to assume that flagella originated in very ancient times, as they provided one decisive advantage: movement.

Looking for food

Some scientists believe that the oceans were seething with food during primeval times; they formed an enormous pot of organic material generated by the processes underway in the atmosphere. In reactions triggered by solar energy and lightning, the primitive gases produced an uninterrupted rain of complex nutrients, which these cells could "attack" and degrade though a process called *fermentation* thanks to enzymes located in their cytoplasm, as still occurs in present-day prokaryotes.

As is known, enzymes are molecules that have the unique capacity to accelerate specific chemical reactions, thereby transforming the matter with which they come into contact. They can be compared to the tools a craftsman uses to transform wood or metal into useful objects such as pliers, a saw, a screwdriver, pincers, and a plane.

Analogously, the enzymes of these primitive cells could have converted the organic molecules floating around in the water into "building blocks" for use in their own metabolism and growth. Strangely enough, it has been discovered that bacteria eat the same compounds that constitute our daily diet: carbohydrates, fats, proteins, and sugars. It may have been that prokaryotic cells, these roaming fermenters, had a rather similar diet.

The limits of growth

Those who believe that these primordial fermenters were the first inhabitants of our planet think that the abundance of food provided by the atmosphere favored an incredible proliferation of cells and apparently boundless growth. But at some point there must have been a crisis because these first prokaryotes were pure consumers. They were unable to produce food and could only eat (reduce) what they found in their paths. This situation could not last for long.

Since the population grew faster than available food, the population explosion prompted by the proliferation of these early prokaryotic cells must have caused a resource problem. This problem—one of "sustainable growth"—is well known on the earth today. No population (whether bacteria or humans) can grow faster than available resources without entering into crisis: if growth is to continue, "new technologies" have to be invented to create new resources.

The advent of photosynthesis

Other scientists claim that there is no proof that bacteria were the first inhabitants of this planet. In their opinion, they appeared later. The oldest sedimentary layers, they say, contain no traces of such prokaryotic fermenters: the oldest unicellular organisms of which we have evidence are *photosynthesizers.* This hypothesis asserts that photosynthesis existed from the beginning of life, that it was, in fact, the spark that ignited life. In this view, the "technology" needed to create resources (food) existed from the very beginning.

Only further research will tell us who is right. But in the meantime, we can try to find out what kind of photosynthesis it was that developed in primordial times. Was it the same kind that we know today? Or a simpler version? Or a different version altogether—a volcanic version perhaps?

Producing food rather than eating it

The landscape from our vantage point atop a huge mass of lava emerging from the water looks Dantesque. On the horizon, the rising sun is an immense fiery ball. A volcano just over a mile away continues to spew dense white smoke into the sky. Only a moment ago, the earth below us was shaken by a tremor. Continual rumblings drown out the lapping of the waves. The coastline is barren and deserted. For as far as the eye can see, there is only sand, rock, and lava.

The foul odor of rotten eggs (that is, hydrogen sulphide) hangs in the air. We dip our hands into the water: it is warm, almost hot. There must be a hot spring below releasing heat and gases. This is the best place to look for new forms of life, and we are going to dive in again to do just that.

Back in the water, the setting is still characterized by volcanic rock. The columns of bubbles of all sizes rising from the sea floor make us feel as if we are swimming in a huge glass of champagne. Zigzagging our way through the inverted shower of bubbles, we slowly make our way down to their source.

The water is hot down here. The rocky sea floor feels unusual to the touch, as though it were covered by a fine layer of slime.

We reduce size to be able to explore this surface, as everything is much clearer at the microscopic level. And indeed, the landscape has changed completely. The hot springs—not only the rocks, but the water—are churning with multitudes of tiny cells intent upon carrying out that minor marvel that allows them to *produce* food and not only to *eat* it.

Playing Lego with hydrogen

These minuscule organisms, now known as *sulphur bacteria,* are capable of producing food for themselves very efficiently from simple chemical substances. It would be as if we were able to form steaks directly inside our stomachs by taking in the right atoms of carbon, hydrogen, and so on—no cattle needed at all.

These bacteria assemble and disassemble molecules as if they were playing Lego. The object of their game, however, is to assemble organic molecules (that is, food), starting from hydrogen (H) and carbon (C). In fact, "CH bonds" are typical of life chemistry and are the basis of innumerable organic compounds. How does this assembly and disassembly take place?

The sulphur bacteria extract carbon from carbon dioxide (dissolved in water) and hydrogen from hydrogen sulphide (that is, from the foul-smelling gas bubbling up from the sea floor). With the carbon and the hydrogen, they then form a number of compounds, including carbohydrates.

But they need considerable energy to detach, reshuffle, and reattach these pieces (just as we require energy to detach and reassemble the Lego pieces). And this is where the ingenious trick adopted by these cells comes in: they use the energy from the sun. The photons coming from the sun penetrate the cell and trigger a photochemical reaction with a pigment, setting off the whole chemical sequence.

The chlorophyll trick

This mechanism may be illustrated with an analogy from vision. Photons penetrate the eye and cause a photochemical reaction on the pigments contained in the cells of the retina. This reaction produces a nervous impulse which is transmitted to the brain.

In prokaryotic cells, the pigment is called chlorophyll. Yes, the famous chlorophyll. This is the beginning of its long evolutionary course: what started out as a simple pigment, incorporated (by accident, it is believed) in some of these cells, was later to become one of the most important factors in the development of life on Earth.

But were sulphur bacteria the real pioneers of photosynthesis? Oceanographer Jack Corliss of the University of Oregon feels that life probably originated close to the ancient edges of the continental plates where enormous quantities of hydrogen-rich gases were emitted.

Other researchers, however, do not go along with this idea. It is true that there was much volcanic activity in primordial times, both on land and at sea. But the amount of hydrogen that could be extracted from hydrogen sulphide was in no way comparable to the quantity of hydrogen available from the water in the oceans.

This, then, is the basis for another hypothesis which suggests that photosynthesis developed directly among cyanobacteria and that sulphur bacteria were simply local adaptations, lacking any evolutionary future.

Apart from the question of which of the two holds the record for being the first organism capable of photosynthesis, cyanobacteria initiated an extraordinary process of transformation and evolution, while sulphur bacteria remained closed in their small smelly world.

Let's go, then, and take a look at these remarkable cells capable of manufacturing their own food from carbon dioxide, water, and sunlight.

The strategy of the cyanobacteria

Only a few yards below the surface of the water, the multitudes of cyanobacteria look like specks of dust in the sun.

In order to gain exposure to sunlight, they live as close to the surface as ultraviolet radiation will permit. As already mentioned, the sun's rays are lethal up to a depth of thirty feet, making this the minimum depth at which cyanobacteria can live (like all forms of life at this time).

The cyanobacteria's specialty is their ability to convert light into energy by splitting the water molecule. This strategy is much more profitable than the one used by sulphur bacteria for one simple reason: the amount of hydrogen that a cell can extract from water (to build or repair itself) is practically unlimited. In fact, water is composed of H_2O, that is, one oxygen atom (O) and two hydrogen atoms (H). This means that these cells do not have to stay close to sulphur springs to obtain hydrogen: the entire ocean is their hydrogen reservoir.

From the pear to the baobab

Inside their soft mucilaginous membranes, the cyanobacteria already contain all the wonders that allow for this kind of photosynthesis, including betacarotene, mixoxanthine, and phycocyanin. They have vacuoles, some RNA, and above all chlorophyll which, together with phycocyanin, gives them their blue-green color. Chlorophyll is still in the form of free-standing opaque sheets and is not yet enclosed in a membrane or stacked into piles as will occur later in the chloroplasts.

In miniature, these cyanobacteria contain all the basic ingredients needed to give rise to the whole range of vegetation found on Earth, from lichens to sequoias.

The entire plant kingdom (including trees, shrubs, flowers, and grasses) exists thanks to the ingenious biological invention of photosynthesis, and much of the animal kingdom as well relies on the food created by photosynthesis (herbivores directly as they eat plants and carnivores indirectly as they eat herbivores). From the pear tree to the baobab, from buffalo to lions, squirrels, insects, and human beings, most of the food chain is dependent on this basic "trick."

As an old philosophy professor used to put it: "The sun is that thing which makes grass grow, so that cattle can eat it, so that they can provide your professor with his steak, so that he can stand here in front of you and talk to you about Plato." This eloquent illustration of the cycle from the sun to the philosopher illustrates the crucial role that photosynthesis plays in life. Its biochemical mechanism is so impeccable that it has remained almost unchanged throughout time and continues to function today almost exactly as it did in primordial times.

The release of oxygen

There is, of course, another aspect to chlorophyll photosynthesis that is extremely important: the release of oxygen, the oxygen which slowly filled the oceans and the atmosphere, making life as we know it today possible on Earth.

Oxygen is basically a waste product of photosynthesis. Once hydrogen has been extracted from water, oxygen molecules are left over, which are eliminated as waste just as exhaust gases are eliminated from the engines of our cars.

As we will see farther on, oxygen was poisonous for primitive organisms. That is precisely why it provoked totally new evolutionary adaptations, leading to the birth of new organisms that not only knew how to defend themselves against this poison, but were able to use it as an extraordinary new source of energy.

We will return to this point, but right now, a few molecules of oxygen have just been released beside us. Let's follow and discover what their destiny will be. But you had better be ready for surprises! These molecules will take us to a very strange place, a kind of cemetery for primordial oxygen which attests to the ancient origins of photosynthesis.

A microscopic alga, a buffalo in the shade of a baobab tree, and a tiny beetle. "The entire plant kingdom exists thanks to the ingenious biological invention of photosynthesis. Not only that, but much of the animal kingdom relies on the food created by photosynthesis (herbivores directly as they eat plants and carnivores indirectly as they eat herbivores). . . ."

The molecules are now grouping into tiny bubbles. We will have to become even smaller—thousands of times smaller—to see them at close hand.

Riding an oxygen molecule

No larger than a millionth of a millimeter, we scramble atop the oxygen molecule that has just been released from the cell. With us astraddle, it starts to float around aimlessly in the sea of atoms, ions, molecules, and macromolecules, like a person lost in a crowd.

Temperature variations and currents have a chaotic effect on these particles and drag us this way and that. Suddenly we are sucked into a tiny whirlpool along with other molecules. Then, as we are whisked along upside down, we feel a slight vibration followed by a jolt: our oxygen molecule has hooked up to something else. We have been involved in a chemical reaction—an oxidation!

The structure of our molecule has changed completely: it is much larger. Other atoms—oxygen, silicon, and iron—have latched onto ours. A total of twelve atoms are now linked together in various ways. The final result looks rather like a bunch of grapes: this is an iron compound.

Our traveling speed has been reduced and our now much heavier molecule starts to sink slowly toward the sea floor. All around us, similar molecules are forming and starting to sink. But where has all the iron required to produce these compounds come from?

The earth's crust contains huge amounts of iron. And since it is constantly being eroded by rains and storms, the water flowing into the ocean carries with it debris of all kinds, including iron particles. When they come into contact with the oxygen produced by the cyanobacteria, the iron particles borne on the currents along with grains of sand form compounds that deposit on the sea floor.

We are now descending more rapidly into the shadows: the water is getting much colder and visibility is decreasing. Our bunch of grapes settles to the bottom with a bump. Around us is an endless expanse of iron compounds, giving the sea floor a reddish brown tinge, the color of rust. Other molecules like ours are still descending, as if in an enormous parachute landing; some come down right on top of us.

In order to avoid being trapped under a layer of iron compounds, we abandon our molecule and struggle toward the surface, only too happy to leave this icy, inhospitable place.

The revealing rust

These layers of "rust" are the oldest traces of the presence of oxygen in the primordial seas. Alternating with fine layers of a grayish color (perhaps the result of seasonal changes), they form "striate (layered) formations," which tell the story of the marine environment from the point of view of oxygen.

The layers are very ancient indeed. Geologists have discovered that they date back as far as 3.5 (or even 3.8) billion years. Other researchers have identified layers four billion years old using other dating methods (based on the percentage of carbon 12 or carbon 13 contained in certain rocks found on the sea floor).

In any case, they are veritable oxygen cemeteries and reveal a very important fact: that photosynthesis already existed at the dawn of life. This means that cells capable of using the sun as a source of energy were already present and active, and that the cyanobacteria may have been the earth's first inhabitants.

An underwater city

The cyanobacteria have more surprises in store for us. Traveling around the sea floor at a depth of approximately thirty feet, we approach an area covered with strange-looking muffin-shaped rocks.

At first glance, the most striking feature is their regular shape: they could hardly be the product of erosion or lava deposits. If they weren't quite so round, they could vaguely resemble termite hills in Africa. A few are sometimes grouped together, creating irregular shapes and branchings.

We drop down to the sea floor to take a closer look at this surprising "petrified forest." This is the first time since our landing on the planet half a billion years ago that we have seen anything of the kind, anything so organized. It almost looks as though someone has planned and built this city, this ghost town without inhabitants or motion of any kind except for the dim rays of light filtering down from above.

Rocks with billions of inhabitants

At our size—a few thousandths of a millimeter—these rock formations look like the Dolomites. The best way to explore them, therefore, is to cruise around at low altitude in the valleys and canyons separating them.

Life immediately becomes evident. The surface of these apparently inanimate objects is teeming with enormous quantities of cyanobacteria. Instead of living alone, these cousins of the cells observed previously have grouped together into kinds of hives, forming a dense, compact surface composed of billions of individuals. They have built these large muffin-shaped rocks by depositing layer upon layer of calcium carbonate, generation after generation. This city has not expanded into the outskirts—the "suburbs," as they say—it has grown by building on top of existing layers.

Slowly, in the silence of the primeval oceans, these cities have organized and grown, spreading everywhere and forming bizarre megalopolises. The ideal environment for them are shallow coastal waters where solar energy is available. In fact, they cannot live on the sea floor because they need the sun, while at the same time, they cannot emerge onto dry land because they have no protection against the ultraviolet radiation.

These complex "settlements" (which sometimes take the shape of flat calcareous deposits) are called "stromatolites."

"Traveling around the sea floor . . . we approach an area covered with strange-looking muffin-shaped rocks."

Some have survived the great changes on this planet and still exist today. They can, however, be found only in certain small bays in which the extremely salty waters produced by evaporation tend to keep their predators (mollusks) away. The cyanobacteria of which they are composed carry out photosynthesis by splitting the water molecule and producing oxygen as a waste product, like their floating cousins.

The surprising thing, here too, is the age of these formations: some stromatolites date back 3.5 billion years. This is just another confirmation of how ancient photosynthesis is.

Although these algae live next door to each other, they do not form colonies: each cell lives independently of its neighbor. Much time will have to go by before communities of cells appear in which each is able to communicate actively with its neighbors and even more will go by before "societies" come into being in which each individual has its own role.

Here we are still at the level of a crowd at the stadium. We will have to wait a long while still before we can witness organization equivalent to that of a polyphonic choir. But the fact that these algae live together may nevertheless have some kind of initial nebulous significance. In the primordial ocean, under the relentless and apparently unchanging cycle of night and day, the slow diversification of life forms has begun. Through mutation and selection, each is contributing to life's ramification into continually different shapes and forms, the whole range of which will probably never be known.

The appearance of this primeval "Stonehenge" and the geometry of its monoliths, which constitute the first constructions on our earth, suggest that it is time for us to stop our exploration of the early oceans. Let's return to the surface and give our story some time to unravel.

See you in a billion years.

MAY 16TH–JULY 1ST (2.5–2 BILLION YEARS AGO)

A billion years later

As we sit on the beach, the sun comes up between two large dark rocks in front of us, casting its shimmering reflection across the immense sea. The landscape is as lunar and desolate as ever: not a blade of grass, not a sound, not a movement on this earth. Total silence.

And yet many things have changed in one billion years. The air has cooled down. The clouds are whiter: the volcanic eruptions which formerly filled the sky with dust and particles have finally diminished. But as a result, the sky no longer has those rich crimson hues at dawn and dusk that we observed when we first arrived.

The geography has also changed: in one billion years, the continents have shifted, collided, repelled, and once again approached each other in a slow but inexorable drift which has redrawn coastlines and landscapes and created folding and mountain chains at the collision points.

The moon is smaller. In fact, with each orbit, it has gradually and imperceptibly moved away from the earth in a slow process that is bound to continue. Those familiar markings are appearing on its surface.

The tides have been affected by the fact that the moon's gravitational field is moving away and are now less intense. Days (and nights) are appreciably longer.

The sea stretching out in front of us is a deeper blue. Seen from space, the earth may look very much like it does today: the quantity of water on the planet has increased (thanks mainly to the action of the volcanoes) and the oceans have taken on the dimensions and colors with which we are all familiar.

The incredible thing is that though we are almost halfway along the road to life on Earth (midsummer on our calendar), the planet is still deserted. Nothing seems to have changed, even under the ocean's surface. The only forms of life are microscopic and primitive prokaryotic cells (that have no nucleus). Nothing else.

And yet something important is in the offing. We'll have to return underwater, however, to understand it better.

New models on display

At cell size, we once again find ourselves surrounded by hordes of microorganisms. And now the difference with respect to one billion years earlier becomes noticeable: unseen forms of life drift past us.

This is like returning to an automobile show after years of absence. Production has been enhanced and diversified; alongside the old models—still functional—there are improved versions as well as totally new models. Although all models are based on the same basic machine, there is now a far wider range. For example, there are rod-shaped, sphere-shaped, and comma-shaped models (the primordial equivalents of bacilli, cocci, and vibri). Some models are arched or spiral-shaped. Some drift on the currents, other swim with the help of a variety of cilia and flagella. There are also numerous variations of the same model (in fact, there are 1,500 different species of cyanobacteria today).

Evolution is obviously underway. Very slowly, but underway nevertheless. It is moving and preparing the ground for the next leap, the one that will give rise to nucleate cells (the cells that form the basis of all protozoans, plants, fungi, and animals that exist today).

As we look at this diversity, an analogy comes to mind: that of a deck of cards. It takes a large number of basic cards to make up a deck, but they allow for innumerable combinations and therefore a wide variety of games.

Things are somewhat the same for DNA: these prokaryotic cells correspond to the individual cards that are differentiating to prepare a kind of basic deck to be used later on in much more elaborate and complex games (we will return to this concept in connection with the birth of nucleate cells).

But why is this differentiation process so slow?

As mentioned, there is already an enormous variety of forms, but it is taking them an extremely long time to change. Looking at our calendar, we are surprised to see that while the first living objects appeared shortly after the formation of our planet, perhaps as early as four billion years ago, the development of more complex cells is moving at a very slow pace indeed.

We have been "on the road" now for approximately two billion years and the prokaryote "automobile" is still substantially the same (even with all the important variations and innovations we have seen). Not only that, it is destined to remain that way for a long time to come (until about August 25th, approximately 1.4 billion years ago), that is, until the appearance of the first more complex cells: cells with a nucleus—eukaryotic cells.

Amazingly, it will take almost two and a half billion years (two-thirds of the entire path toward life) to pass from a prokaryotic cell to an eukaryotic cell and only 1.4 billion years to pass from an eukaryotic cell to a human being.

The long "deadlock" in life

Numerous hypotheses have been put forward to explain why evolution (which sometimes goes through periods of strong acceleration) seems to have entered a period of deadlock in this case.

The phenomenon may appear all the more surprising in that prokaryotic cells duplicate very quickly and, for this reason, seem particularly suited to undergoing rapid evolution through the well-known mechanism of accidental mutation and subsequent selection by the environment.

It may be helpful to go over this mechanism briefly: due to the action of mutagenic agents (radiation, chemicals, etc.) or errors in duplication, DNA occasionally undergoes minor chemical changes resulting in the production of altered proteins.

These changes (mutations) generate anomalies that are generally lethal for the individual. In some cases, however, they may prove either useful or irrelevant for the survival of the organism. These useful or neutral changes are therefore passed on in successive duplications of the DNA, ready to become dominant whenever the environment gives them a chance to assert themselves.

Studies of bacterial cells have shown that *Escherichia coli,* for example, a well-known bacterium used in many experiments, undergoes duplication every twenty minutes. This has made it possible to estimate the rate at which random mutations take place: mutant cells account for approximately 1.5 percent of the total cell population after thirty cell duplication cycles (and thirty duplications take only ten hours!).

Just think what could have happened in the course of billions of years, with the continual duplication of billions of billions of individual organisms. And yet, there was no evolutionary leap even vaguely comparable to that which came later. Why not?

A few hypotheses

There is no accepted explanation, only hypotheses. For example, some researchers feel that one of the decisive factors was the relative stability of the environment. Evolution is, in fact, favored and accelerated by the occurrence of profound environmental changes which, through selection, facilitate the emergence of new forms of life possessing better fitness. In this way, certain small groups of organisms endowed with some particular property suddenly have a selective advantage and start to spread, opening up new evolutionary trends (more or less the case of dinosaurs and mammals). Under steady environmental conditions, this is unlikely to occur.

One of the principal factors in evolutionary acceleration is definitely oxygen, which in due course gave rise to a new kind of metabolism. But the oxygen concentration in this initial period is still too low, and it will take even longer before an ozone shield forms to protect organisms from ultraviolet light, allowing them to emerge onto dry land.

Other factors must therefore be sought to explain the evolutionary sluggishness of prokaryotic cells. One is notably the fact that the mutations affecting prokaryotes take place in relatively simple and "closed" systems. That is, they involve rather rudimentary DNA chains and, above all, occur in systems lacking communication, that is, sexuality (as far as we know). Prokaryotic cells merely duplicate themselves, generating twins. If we were to do the same, our children and grandchildren would be carbon copies of ourselves and we would probably have to wait thousands of years to see an individual with even slightly different features.

Sexuality, on the other hand, allows for an amazing amount of genetic recombination, which makes it possible to generate individuals that are constantly different and therefore more suited to new evolutionary scenarios. In fact, sexual reproduction introduces the advantage of the coupling of chromosomes and the appearance of dominant and recessive traits.

But is it conceivable that in billions of years of evolution, prokaryotic cells did not come up with some kind of sexuality or at least genetic recombination?

The dawn of sexuality

We know absolutely nothing in this regard. All we do know is that some present-day bacteria possess a rudimentary form of sexuality known as "conjugation." During conjugation, two individuals unite and one of the two transfers genetic material to the other.

This phenomenon was first observed under the microscope in 1946. Until that time, it was thought that bacteria reproduced by simple fission. Instead, researchers found that there are donors and receivers (one is tempted to say "males" and "females") and that certain donors are more numerous than others.

Another type of DNA transfer from one individual to another is also known, but it has nothing to do with sexuality: it might be called "genetic cannibalism." Certain bacteria feed on the remains of dead cells and ingest their DNA, which then comes into contact with the internal DNA of the cell. In this very elementary way, the ultimate aim of cannibalism—to incorporate the qualities of the dead being—is achieved.

To be quite honest, no one knows exactly how the prokaryotic cells that lived in the oceans billions of years ago behaved. We have no idea whether they practiced some form of sexuality or cannibalism, or whether the properties observed in contemporary bacteria emerged at a much later stage. All we know is that it took life a long time to prepare for the next step.

As we mentioned earlier, however, at a certain point a new element—oxygen—came onto the scene, causing a revolutionary transformation in the environment.

Released in ever increasing quantities by the cyanobacteria into the water and, later, the atmosphere, oxygen destabilized the environment. Somehow oxygen acted like a new "predator," imposing new adaptations and triggering an unpredictable evolutionary spurt. In fact, this "predator" generated respiration as a defense, giving rise to a long line of living creatures that extends all the way to human beings.

So let's go back one last time to our cyanobacteria to see what kind of havoc they have caused with their photosynthesis.

AUGUST 24TH (1.4 BILLION YEARS AGO)

Oxygen poisoning

A quick look at the seas reveals that oxygen molecules are increasing rapidly. The cyanobacteria are continuing to release oxygen as a waste product and the pure oxygen is polluting the environment.

As strange as it may seem to us, since we are so used to considering oxygen the symbol of life (in the absence of oxygen our organisms die in a few minutes), this analogy actually reflects reality quite well: the environment is being poisoned by "noxious" oxygen emissions. For the primitive cells of these primeval oceans, it acts as a potent poison which is destroying them by oxidization. And this situation is destined to have overwhelming consequences for life as a whole.

Oxygen is also slowly invading the atmosphere. There are only slight traces of it now, but the concentration of this poisonous gas will gradually increase. Eventually, it will change the composition of the air and form the famous ozone shield at higher altitudes, filtering ultraviolet radiation (and thus preparing the world for the emergence of life on dry land).

But that is still far in the future. In the meantime, drastic changes are taking place down there, underwater. The massive presence of oxygen is causing the disappearance of those cells that are unable to cope with the new situation. At the same time, it has given rise to a new kind of cell, one capable not only of coexisting with oxygen but also of using it as fuel—an "aerobic" cell.

A masterstroke

This marvel was accomplished in a very simple and clever way. Aerobic cells originally belonged to the line of the "fermenters" (that is, those primordial prokaryotes which fed on organic matter present in water); the trick is that they have learned to oxidize the leftovers of fermentation. That is, they now use oxygen molecules to recycle waste products, causing a new chemical reaction. They are "turbo" cells, so to speak, that use exhaust gases for further "combustion."

These microorganisms have also developed the ability to procure oxygen for themselves by using their enzymes to break down oxygen compounds. In this way, aerobic cells draw new energy from organic wastes: eighteen times the energy that could be produced from the same quantity of raw material by fermentation alone. A masterstroke!

This extraordinary biochemical "invention" is the first step along the long road toward respiration, which is to be so successful in the history of evolution. Although still extremely simple, the aerobic cell is basically the first living creature to "breathe," in the sense that it uses oxygen for its vital processes.

What is happening to the cells that could only ferment? They are withdrawing to oxygen-free places where a supply of organic matter is still available: the muddy sea floor and other places which we "aerobic" creatures consider inhospitable (areas with high salt concentration or hot or acidic springs).

Today, almost one and a half billion years later, their descendants are still tucked away in these marginal environments: rather like the Japanese soldiers who withdrew into the forests and caves of the Pacific islands awaiting liberation, these cells lost the war against oxygen and withdrew to their "niches." They are now referred to as "anaerobic" bacteria, that is, bacteria that live without air. They play an important role in the breakdown of organic material and can be found, among other places, in lake bottoms, the digestive tracts of animals, and thermal springs.

And what about the cyanobacteria? These organisms very soon learned how to live with the oxygen they produced. As a result, they continue to live and to use solar energy for photosynthesis.

Three models of life

Based on their relation to oxygen, three very different models of life are emerging from these oceans:

1. *obligate fermenters,* which are increasingly withdrawing to places uncontaminated by oxygen;
2. *oxygen producers,* such as the cyanobacteria, which generate oxygen through photosynthesis;
3. *oxygen consumers,* that is, the new aerobic cells, which use oxygen as a source of energy.

Actually, these very ancient "models" are the precursors of the three great models that we still find in nature today: fungi, plants, and animals. Fungi ferment organic matter and do not use oxygen, plants produce (and use) oxygen, and animals use oxygen for energy but are unable to produce it. Three models based respectively on fermentation, photosynthesis, and respiration.

Of course, there are intermediate and composite forms. For example, there are bacteria today that are both aerobic and anaerobic; blue algae that can ferment; and plants, such as lichens, which are practically fungi but have trapped an alga to carry out photosynthesis. But above and beyond the variety of living beings found on the earth today, the relationship with oxygen is still a "birthmark" which arose and developed in the primeval oceans.

Another "birthmark" which developed at the same time was the relationship with food. The distinction that we have made between "consumer" and "producer," that is, between organisms that "eat" organic matter and others, like the cyanobacteria, which produce their own food, can be found throughout life. This distinction will give rise to two great kingdoms, the animal kingdom and the plant kingdom. Or in technical terms, it will give rise to "heterotrophic" beings, that is, animals which require other beings in order to live, and "autotrophic" beings, which depend only on themselves for their growth.

Toward growing complexity

The scene has almost been set for the advent of new protagonists that will receive and transform the genetic heritage accumulated in the oceans and usher in a new era in the history of life on Earth.

Most of the preparations have been completed: inorganic matter has given way to living beings capable of movement, growth, reproduction, and respiration. From this point on, further evolution will be directed toward greater complexity, that is, attempts (in various directions) to assemble and reelaborate existing elements in order to construct, first, more complex cells and then pluricellular organisms.

The first part of our journey to the primordial oceans has been longer and more complex than predicted. We hope that those who have had the determination to follow us have managed to get through the bogs we encountered along the way. But this initial part is fundamental in understanding how life's alphabet was set up and how the first syllables and words were formed. This will allow us to understand how languages developed later on.

Of course, not everything is clear. As we have so often repeated, we are only at the dawn of discovery. We can only read the headlines, a few subtitles, and occasional fragments of the story of life. But thanks to the work of a host of researchers, we are now beginning to understand the main lines of the plot. And future work will certainly facilitate our reading even more.

Let's set off now on a new exploratory journey in an attempt to understand how life's tendency toward greater complexity evolved.

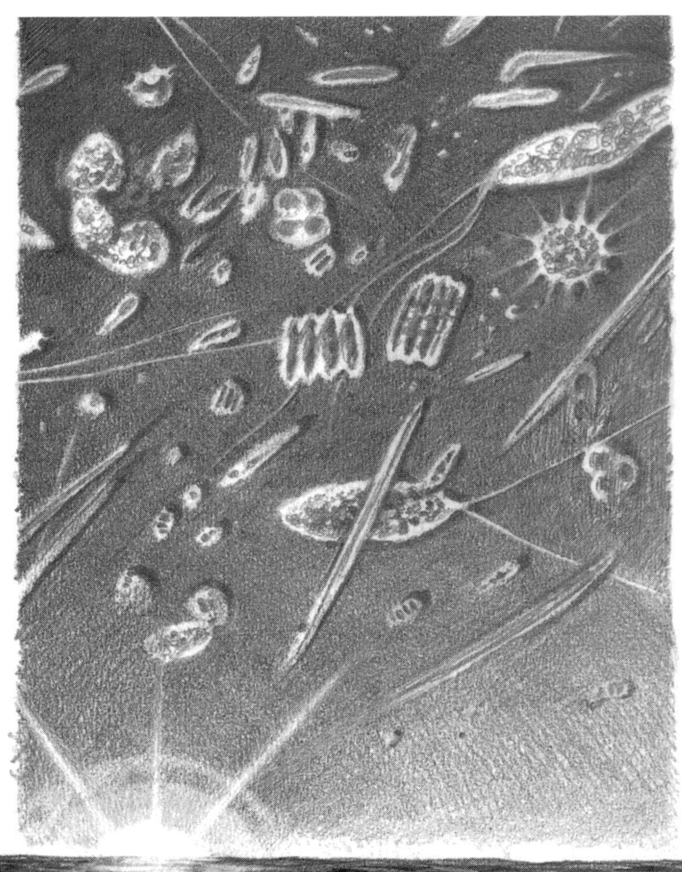

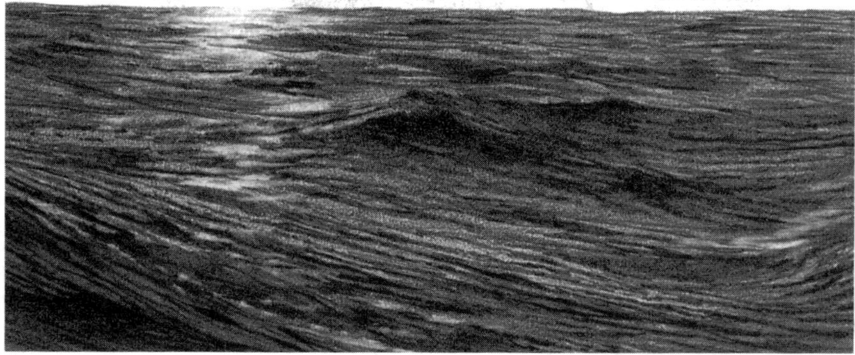

"Things are looking quite different underwater these days. The seas are full of extravagant living forms floating around like hot-air balloons."

3

Toward Complexity

AUGUST 25TH TO SEPTEMBER 10TH (1.4-1.2 BILLION YEARS AGO)

Oceans instead of volcanoes

From our rocky perch, we watch as the sea washes up onto a long stretch of coastline. Four hundred million years ago there was neither ocean nor coastline here; only a vast expanse of desert and some very active volcanoes.

It was the great underground fractures in the earth's crust (faults) which branch out in this area that favored this enormous cleft and, over a period of millions of years, the creation of two immense pieces of land. These two pieces are now separating like a loaf of bread being torn apart: in the distance we can see the shores of what will become a new continent as soon as the last tongues of land have split.

Behind us, a river winds down to the ocean, eating its way

into the terrain. At one point, it cascades into a magnificent waterfall. What a strange feeling: everything looks familiar (the coast, the ocean, the rivers, the valleys) and yet different. The absence of vegetation, especially along the river, produces an unreal atmosphere, as if this were not our earth.

The colors of the sky, the clouds, and the sea are not quite as we know them: everything is slightly tinged with violet. The oxygen that is beginning to accumulate in the atmosphere will soon enrich the general tones of this picture. In fact, so much oxygen is being produced by the photosynthesis of microorganisms (cyanobacteria) that it is gradually escaping the ocean and starting to spread into the atmosphere.

Even the iron which originally trapped the oxygen and precipitated it into the red layers at the ocean bottom can no longer counter this growing production.

It will take a long time before an appreciable amount of oxygen accumulates in the atmosphere: today, on August 25th, its concentration is approximately 1 percent (but this is already enough to block a large fraction of the ultraviolet radiation). In October it will be around 7 percent and in November it will reach 10 percent. From that point on, the ozone layer in the upper reaches will allow life—which was segregated underwater for billions of years—to start to emerge and conquer dry land. The oxygen concentration will eventually peak and remain constant at about 20 percent.

By the time it leaves the water, however, life will already be highly developed with well-structured organisms that are still unthinkable at the present stage. Their development will be made possible by a fundamental evolutionary innovation: the appearance of complex cells with organelles and, above all, a nucleus containing DNA. These cells are called *eukaryotes* (from the Greek *eu*=well, and *carion*=nucleate).

Let's dive into the ocean again to take a look at them. We'll have to scale down to their size, however: no more than a few tenths of a thousandth of a millimeter this time.

Dirigibles, hats, and flasks

Things are looking quite different underwater these days. The oceans are full of extravagant living forms floating around like tiny hot-air balloons. They are around ten times larger than the prokaryotic (bacteria-like) cells encountered previously. They also live much closer to the surface of the water: the oxygen in the atmosphere is starting to filter the ultraviolet light, making it less harmful. Life is no longer relegated to depths of at least ten yards in protected coastal areas; microorganisms can now spread to the open seas without fear of being borne to the surface by the currents. New horizons are opening for life.

We pass an organism that resembles a dirigible and has two mobile flagella at the rear end that act like a screw; smaller cilia at the center stabilize movement. We are reminded of a helicopter, which also has a smaller rotor with the same stabilizing function.

Another organism now floats by. It is slightly rounder, has no flagella and moves in a very bizarre way, like a sack that changes shape all the time. Right now we see a protuberance forming on one side. No, there are two protuberances. It looks as though the organism is growing tentacles.

On the left, a fast-moving, sausage-shaped form spirals by at an incredible speed. This is a microscopic zoo!

Descending another few yards to the muddy bottom—and doing our best not to get stuck in it—we come upon even more incredible creatures. Agape, we watch as they zip, slither, and crawl by. Some of these protozoa swim; others are attached to the bottom like upside-down glasses; still others look like miniature dirigibles, flying saucers, cacti, hats, or flasks with tentacles. Have we landed on another planet? No science-fiction author could possibly imagine such a variety of fantastic creatures.

Now a transparent protozoan passes by. It almost looks like

a plastic container putting its "innards" on display. And those "innards" certainly catch our attention. The most noticeable thing is a small dark mass at the center: the nucleus. This nucleus contains the DNA that was scattered throughout the cytoplasm of the bacteria-type prokaryote cells. The DNA, now much longer than before (approximately one thousand times as long) is coiled around itself. During cell division, it will duplicate itself, forming two identical nuclei, which will contain the genetic material of the two new cells.

Various other "shadows" can also be seen: these are the organelles. The whole thing looks like a motor with all its inner workings exposed: visible are the *vacuoles* (i.e., vesicles that have different functions in different protozoa, such as pumping, storage, and digestion); the *ribosomes* (for the synthesis of proteins as directed by DNA); the *endoplasmic reticulum* (a complex system of ramified tubes); and the *mitochondria* (strange organelles that have their own DNA and provide biochemical energy for cell functions).

The extra-nuclear DNA possessed by the mitochondria is a very interesting anomaly. In fact, many scientists believe (as we will see later) that mitochondria are "stowaways," that is, that they are ancient bacteria that in the course of evolution gradually installed themselves inside other cells through a process of symbiosis and have now become a regular cell component.

These and other organelles are now a part of all modern cells, including those of human beings. In fact, the cells floating around here are already similar to modern cells in every way. But this is where our observation of the microorganism comes to an end; the protozoan starts to swim away, its cilia moving in a synchronized fashion like the oars of a Roman warship.

But where have all these microorganisms come from and how have they formed? What evolutionary course have they taken?

Cells and Company

Unfortunately we lack detailed information here, too. There are, obviously, almost no fossil remains from this period, as cells do not fossilize as bones do. Only a few fortuitous "imprints" provide sporadic and tenuous traces. To date, approximately eight thousand fossil imprints of cells have been found in various localities. An interesting fact that can be gleaned from them is that there were only bacteria-type (prokaryotic) cells up to 1.5 billion years ago; then suddenly, much larger, more complex cells (the ones we are talking about now) appeared.

Of course, a statistical sample of eight thousand findings is too small to provide reliable results (suffice it to think that millions of microorganisms can exist in one cubic millimeter). And, as always, science avoids offering explanations that are not based on sufficient evidence. All it can do is gather data and formulate hypotheses.

One hypothesis is that certain archaic cells (like bacteria) soon began to become more complex, giving rise to an evolutionary line that led to increasingly organized cells.

Another, very intriguing hypothesis, however, is based on the idea that there was a totally independent line which provided an evolutionary "shortcut": symbiosis.

The game of Lego continues to provide a fitting analogy, only this time the blocks are not single atoms or molecules, but entire cells, assembled to form more complex systems.

This idea, first advanced at the turn of the twentieth century, has been taken up again recently and developed by Lynn Margulis, an evolutionary geneticist at Boston University. According to Margulis, modern eukaryotes are the result of the association of a prokaryote capable of using oxygen to produce energy by "burning" sugars with a primitive fermentative eukaryote (or nucleated cell) unable to use oxygen in this way. Together, the two cells became capable of extracting sugar from the environment, on the one hand, and of using it as fuel to produce energy on the other. It is rather like the old story of the blind man and the cripple: their association worked to the advantage of both.

An eloquent example of that kind of association exists in nature today: lichens. As mentioned earlier, lichens are formed through an alliance between an alga and a fungus: the alga provides the system with energy; the fungus provides a dwelling place. Together, they can adapt much better to the environment than they would be able to do individually.

Scientists studying lichens have shown that if separated in the lab, the alga and the fungus can survive independently. Thus they are two distinct forms. But in nature, the two "partners" automatically associate, even during reproduction. Therefore, the association (or symbiosis) is not only a matter of convenience, as it may have been for certain forms in the past (or as it is still today for the pilot fish and the shark), but a permanent association which is perpetuated through reproduction.

Similar associations with other primitive bacteria endowed with chlorophyll (i.e., the cyanobacteria capable of photosynthesis) may have made possible the appearance of more complex cells, such as those found in modern plants. In short, assembly was going on everywhere, with evolution linking various pieces and creating new structures.

To support this theory, Margulis recalls that mitochondria (that is, organelles that have their own DNA and that play an essential role in producing energy for the functioning of the cell) are present in almost all modern eukaryotic cells. The idea that mitochondria and chloroplasts derive from ancient bacteria is a well-documented fact.

Much less accepted is the idea put forward by Margulis that union with another primitive bacterium, the spirochete (a small whip-shaped cell with a flagellum), gave rise to the cell's mobility. To support her point of view, she points out that some spirochetes live alone, others live close to other cells (feeding on their surfaces), and still others live permanently attached to other cells. In some cases, multitudes of spirochetes (such as the *Myxotricha paradoxa*) work in perfect coordination to move a cell.

Hopefully we will be able to learn more about this in the future so that light can be shed on the way things really went. For now, all we know is that a great variety of unicellular life is developing in these oceans—forms of life that will dominate for a long time to come.

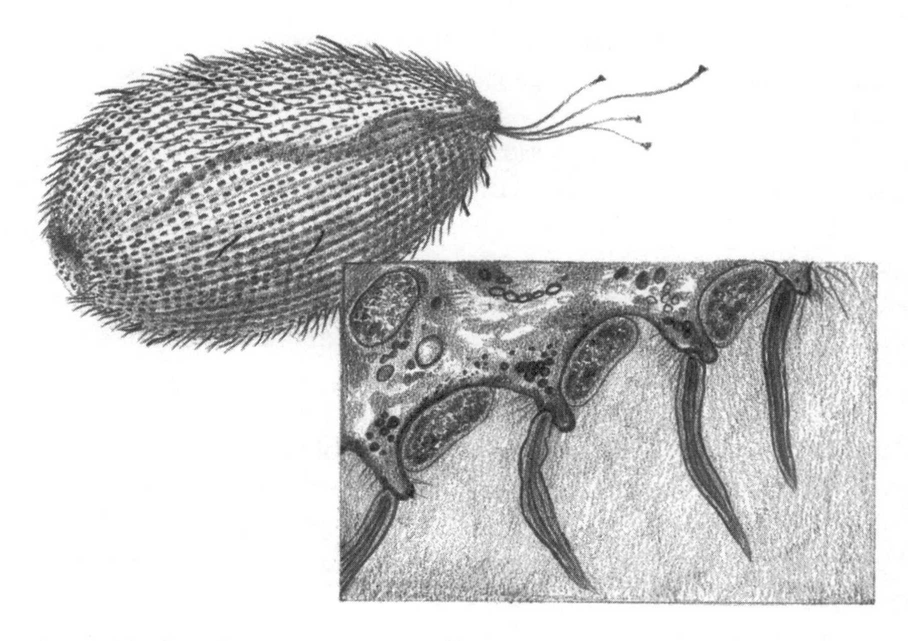

Myxotricha paradoxa. Top: The organism as a whole. Bottom: A magnification of the bacteria that provide locomotion.

SEPTEMBER 28TH–OCTOBER 16TH (1 BILLION–800 MILLION YEARS AGO)

A still apparently lifeless planet

A couple of hundred million years have passed. We are now strolling along a rocky beach not far from our "observatory," a well-equipped tent pitched at the top of a hill.

It's a hot day and huge clouds are scuttling across the sky. Fortunately the wind has died down; for days we were shut up in our little refuge because of an intense sand storm which filled every nook and cranny with sand: our pockets, our shirts, our shoes, our equipment.

Coming back periodically to these waters, we have witnessed the slow but continual evolution of protozoa, still the undisputed masters of these oceans, or, more correctly, of the entire planet.

Indeed, if seen by a passing astronaut, this planet would still seem totally lifeless. No changes are observable on dry land. Certainly, mountain chains have been shaped and leveled; volcanoes have appeared and disappeared; the continents have gone their separate ways, creating oceans in some places and colliding in others. But not a blade of grass has yet appeared and no form of life has emerged above water level.

The oxygen concentration is approximately one-third of what it is today. We can see the molecules of free oxygen (O_2) rising from the ocean to create ozone molecules in the atmosphere (O_3). But the shield against ultraviolet radiation is still not effective and life still has to take refuge under a protective layer of water.

It now becomes clear to us to what extent the evolution of life was (and still is) affected not only by the terrestrial environment, but also by extraterrestrial factors: ultraviolet radiation, the tides, the sun (light and heat), the asteroids, solar winds (which swept away the primordial atmosphere), meteorite fragments, perhaps even comets (which brought not only water but probably the ingredients for a new atmosphere), gravity, and even the kind of orbit the earth makes around the sun. Consequently, the oceans in which this evolution is taking place must not be viewed as a closed system, as a self-sustaining aquarium.

Looking at the ocean, the land, and the sky from our vantage point, we are well aware that everything interacts in a continually changing equilibrium. Even life, after having undergone external influences for so long, is starting to modify the environment through the release of oxygen into the atmosphere.

So what, then, has happened in the oceans in this time?

Strange wandering creatures

During our various dives into the depths, we have noticed some truly extraordinary creatures. For example, we saw a strange cell which, when cut in half, regenerates the missing half; protozoa that are beginning to act like real predators; and other organisms, which, like hyenas or jackals, eat the remains of dead cells.

Some microorganisms seem attracted by the movements of prey (or perhaps the vibrations of the water around them). Others are drawn by "odors" or, rather, traces of chemical compounds emitted into the water.

We have noticed that heat strongly affects many microorganisms, influencing their metabolism and, hence, their growth. Some protozoa seem to respond in a special way to luminous stimuli: in some cases they are attracted to it, in others—if the light is too strong—they are repelled.

In any case, all sorts of novelties are arising down there. But it is impossible to have a complete picture. During the millions of years that have passed, life has spread everywhere—to the oceans, the rivers, the lakes—and has diversified incredibly.

Trying to describe the variety of life populating the earth in this period would be like trying to describe the fauna in Africa by observing only one limited area. How can this problem be overcome?

One way is to do what biologists do: observe under the microscope the extraordinary and varied fauna that have descended directly from primitive protozoa.

In fact, the microscope can be used as a kind of "time machine" to study archaic forms that still live in the same environments and are probably very similar to their ancestors. These are the famous passengers who got off the train of evolution at the first stop and who now provide indirect evidence of a bygone era. These organisms are, of course, different from their ancient ancestors—even if they have maintained many of the same characteristics—as they have been subject to evolution for a very long time. But observing the huge crowds of protozoa moving around in a drop of water under the microscope may nevertheless give an idea of the many different attempts made and the roads taken by evolution in the primordial seas.

Let's give it a try then: a series of slides will provide us with a firsthand view of the distant descendants of the fantastic zoo that lived in the planet's primordial oceans.

A fantastic zoo under the microscope

The first creature we come upon under the microscope is a strange organism that moves in a most unusual manner. It looks like a plastic bag full of water rolling slowly down a staircase. It's an amoeba.

It changes shape constantly, taking on the most bizarre forms. It almost looks like the shadows we make when we move our hands and fingers in front of a projector. In fact, fingers of various shapes and sizes sometimes protrude from the amoeba; they are known as "digitiform processes."

Like a tiny science-fiction monster, the amoeba moves along slowly. Probably attracted by some chemical substance, it approaches a prey: a minuscule single-celled organism that is totally oblivious of the danger. The amoeba's mobile "tentacles" prepare to encircle it in a lethal embrace that will end with the tiny protozoan's death.

The amoeba has stretched out into an elongated, circular shape: it looks like a piece of Plasticine being modeled and remodeled. The prey is surrounded; it cannot escape. The amoeba draws in closer and closer.

At this point, the lethal chemical substances produced by special vacuoles within the amoeba come into play; although they serve for digestion, they may in some way be considered poisons.

The protozoan is disappearing, it has been incorporated by its assailant. Unwittingly, we have witnessed a minuscule tragedy in a drop of water.

A very intense metabolism is now developing inside the amoeba: oxygen, destined to fuel the energy processes, enters the cell through the membrane. Digestion calls for the activation of numerous internal mechanisms and will conclude (as in all living beings) with the expulsion of waste: in this case, carbon dioxide and urea.

The first armored beings

Let's change slides. Here is another primordial creature, armored like a medieval warrior, but with strange "arms" or filaments protruding from the armor. Using a rather less poetic analogy, we might say it looks like a potato masher. The filaments are pseudopods, that is, elongations that the protozoan uses to feed and to move.

This is just one of the many foraminiferans that still populate our oceans today. Foraminiferans come in all shapes and sizes: tubular, spherical, spiral-shaped. Eighteen thousand species are known.

The sea bottom today is composed of layers of their shells or *tests*; such deposits cover an area of 70 million square miles. One curious note: study of the huge blocks used to construct the pyramids in Egypt has revealed that they are made of foraminiferan deposits.

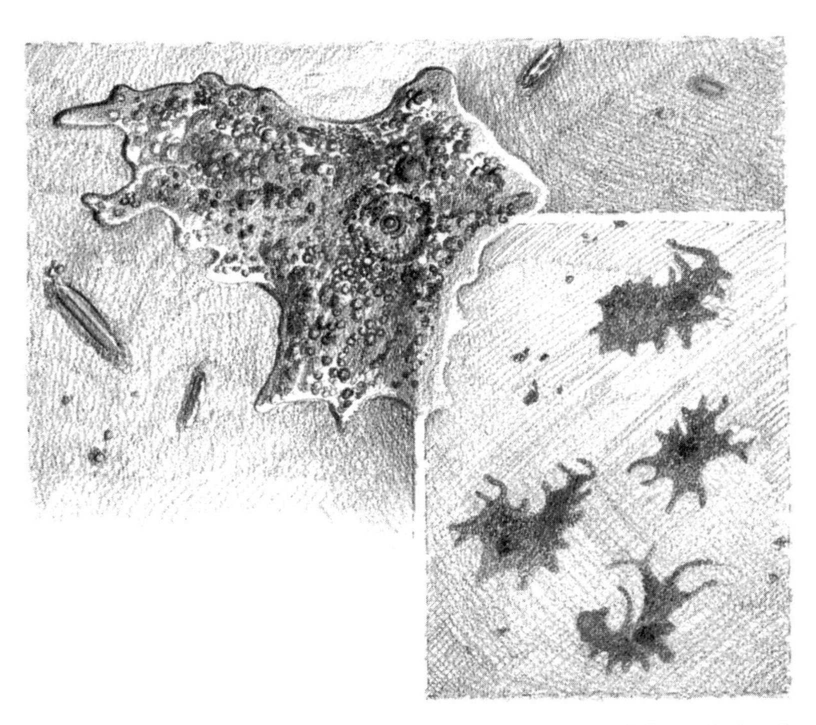

Top: An amoeba, strongly magnified. Bottom: Some of the infinite shapes taken on by these microorganisms during movement.

Tests are a biological invention that will lead to innumerable variations and will meet with great success through time. Providing rigid protection, they may be of different kinds: calcareous, chitinous, and gelatinous (or they may even be made of sand detritus).

Behind these "armored" protozoa are others with a different kind of armor: the radiolarians. The shapes of their tests are absolutely fantastic: some look like the dome of a basilica, others like a spaceship, others like a hollow golf ball. Further behind them is a heliozoan, its filaments extending in all directions to create a sunburst effect.

Suddenly a creature that is "motorized" as well as "armored" starts to make its way across the slide. This protozoan has a very light test and flagella that allow it to move freely. Its flagella are starting to be very specialized: they can serve not only for locomotion, but also for sensory reception and the capture of food.

A second unsuspected planet

It's strange to think that human beings were unaware of the existence of this extraordinary universe for thousands of years for the simple reason that they couldn't see it. These protozoa were being touched, eaten, and drunk without anyone even suspecting their existence.

Only in 1675, with the invention of the microscope, did the human eye suddenly discover this second invisible planet of miniature creatures; protozoa were named "infusers" because they were found in plant infusions.

In those times, ideas about the origins of life were rather confused: it was thought that these life forms sprang spontaneously from the putrefaction of plant matter. Indeed, in the sixteenth century, there were still people who thought that rats could be generated by piles of rubble and rags! Only with the birth of mi-

The extraordinary and fantastic shapes of radiolarian shells.

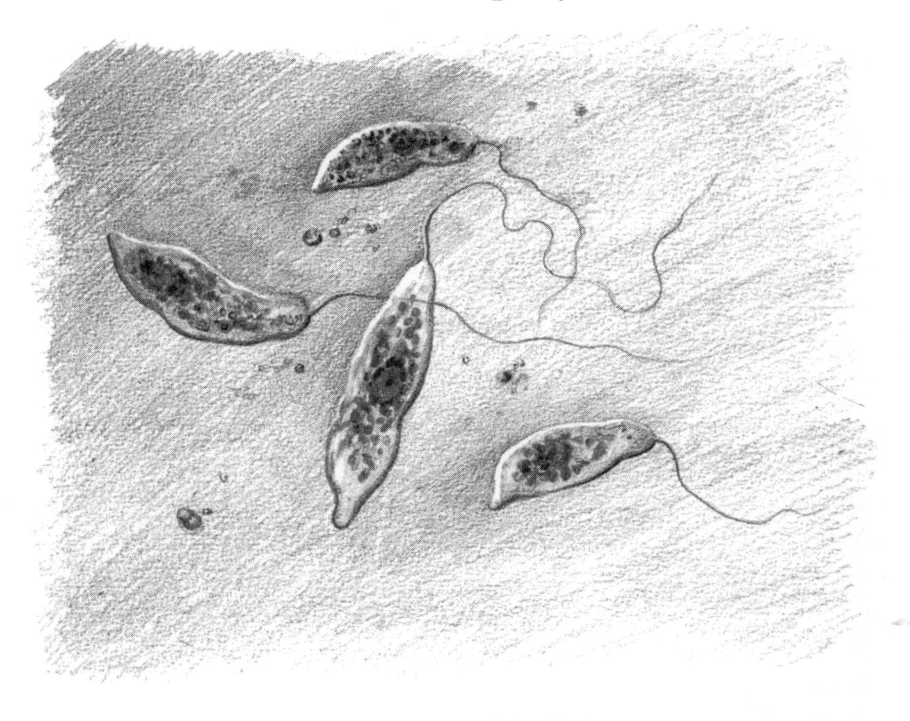

"And yet, the *Euglena* is a very particular microorganism. It is halfway between a plant and an animal. Or rather, it can, depending on the circumstances, behave and feed like either an animal or a plant. Or even a fungus."

crobiology in the nineteenth century did scientists begin to gain deeper insight into this new world and to understand that bacteria and protozoa represent a very primitive stage—a primordial phase—of life on Earth.

One of the first to understand this was the naturalist Charles Darwin (1809–1882). With the intuition of a genius, he even described what he felt to be the initial stage of biochemical evolution, which was to lead to the most elementary forms of life. Prophetically, in 1871 Darwin mused about a pool of hot water containing all the salts of ammonia and phosphorous as well as the light, heat, electrical energy, and so on, needed to form a proteic compound, ready to undergo even more complex changes. This was an amazingly modern view, anticipating and suggesting the basic mechanism that is most probably behind the initial phases of the evolution of life.

Half animal, half plant

We insert another slide. In the midst of a crowd of protozoa is a very special creature. At first glance it does not look particularly unique: rather round and elongated, it is propelled forward by flagella that give it spiral-like motion. And yet the *Euglena* (that is its name) is a very peculiar microorganism: it is halfway between an animal and a plant. Or rather, it is a creature which, depending on circumstances, can behave and feed like either an animal or a plant. Or even a fungus. In a word, it is multifunctional, like powerplants that can run on gas, oil, or coal.

The *Euglena* can capture tiny organisms like a predatory animal (even though this rarely happens); it can produce its own food through photosynthesis like a plant; and if necessary it can use degraded organic matter like a fungus. Not only that, it can also encyst itself. That is, in unfavorable environmental conditions (where food is lacking), the *Euglena* dehydrates and encloses itself within a resistant capsule. By entering into a latent state, it can suspend its vital functions for some time. This kind of encystment can also be stimulated by conditions that are particularly favorable to the activity of chlorophyll, that is, intense light.

Basically, the *Euglena* has a choice of options in different circumstances, but it generally prefers to be a plant when there is enough light, in which case it loses its flagellum. A plant does not have to move; in fact, it would be a waste of energy.)

Under the microscope, the *Euglena* continues its spiral

course, agitating its flagellum in the drop of water. But it suddenly turns toward the light, showing us its "mouth": a funnel-shaped orifice through which it ingests its food. The orifice terminates in a vacuole, that is, a tiny sac linked to others which convey food inward. A primeval digestive system, so to speak, which coexists with the chlorophyll pigments typical of plants.

In its eternal Lego game, evolution has assembled a mixed, multipurpose creature, rather like those fancy automobiles in James Bond movies that can function like motorboats when they are in the water but can also take to the air like planes. Analogously, the *Euglena* can be seen either as a mobile plant or as an animal capable of photosynthesis.

It has changed direction once again and is now moving away. Its movements are not random, however; they are regulated by a primordial "eye" (even if the term is inappropriate in this case). The exact name for this area that senses light located above the mouth is *macula oculare* (or "ocular spot"). The macula enables the *Euglena* to move toward light (as plants do), as light is essential for photosynthesis. But when the source of light is too direct, it triggers the opposite reaction: avoidance.

That is why the light from the microscope initially attracted the *Euglena*, while its intensity now turns it away. In fact, the microorganism veers off behind some foraminiferans and disappears.

The birth of death

What is this fantastic spherical creature that has just appeared? Compared to the other protozoa, it is gigantic: it measures about half a millimeter—about the size of a four-story building compared to a human being.

The elegant sphere is, in fact, a "condominium," a colony

composed of between eight and seventeen thousand perfectly aligned individuals. The interior is hollow (but filled with a gelatin-like liquid which shapes the structure); the grid of flagellated protozoa on the surface makes the whole thing look hairy. A collective "stroke" of the flagella moves the colony forward through the water. This beauty is called *Volvox*. It brings to mind the spherical dandelion flower, which sends hundreds of tiny parachutes into the air when we blow on it.

On closer examination, certain spots on the surface turn out to be concentrations of individuals. One of these spots is pulling away to make a go of it alone. What we are witnessing is probably one of the oldest forms of sexuality. The microorganisms that have pulled away are about to turn into sexual cells (or gametes): the larger ones will become female gametes (eggs), while the others will become the male gametes. They will give rise to new organisms, while the nonreproductive cells (that is, the small flagellates left on the surface of the large sphere after the departure of the gametes) are destined to die.

Thus the *Volvox* clearly represents a major evolutionary step: the birth of death. The Volvox demonstrates that when individuals start to live together, they also start to divide tasks, with some cells specializing in reproduction (with eggs and sperm cells, as in human beings), and others specializing in other functions.

A transition has taken place from minute solitary and self-sufficient individuals (which reproduce individually) to social individuals which associate to build larger structures and which delegate reproduction to particular individuals (or, rather, certain specialized cells). In evolution, this step has meant the advent of death: the efficiency of the whole demands the death during reproduction of the nonspecialized individuals.

All living forms that appeared on Earth in the first two to three billion years (and which we encountered during the first part of our journey) were "immortal," in the sense that each individual produced an exact copy of itself through duplication. This process went on ad infinitum, making it difficult in this society of carbon copies to distinguish the original from the progeny. With the birth of colonies, individuals (cells) appeared that no longer reproduced: at the end of their life cycle, they died, without having generated a copy of themselves.

The *Volvox* seems to tell us this story as it floats off like a bubble. But at what point in evolution did life pass from individualism to sodality? That is, from organisms that live alone to creatures that thrive in groups?

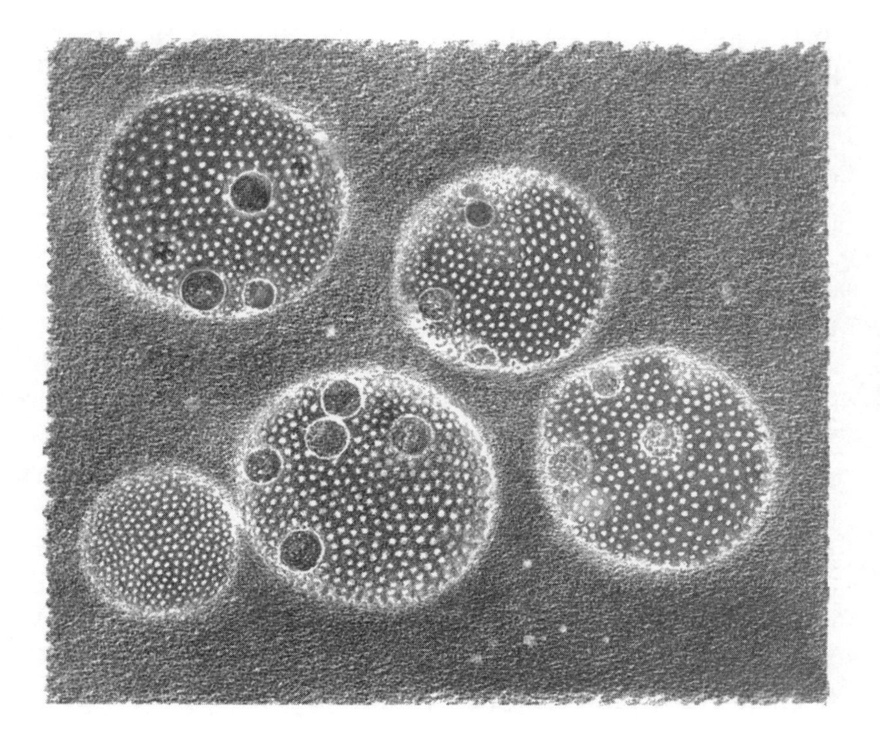

The *Volvox* are spherical colonies composed of thousands of individuals perfectly aligned beside each other.

The first colonies

The story is extremely difficult to reconstruct, but it is fundamental in understanding the transition toward modern organisms. All we know is that the first groupings are very ancient: they date back to the earliest periods of life on Earth.

In fact, prokaryotic cells that lived together—the stromatolites—already existed 3.5 billion years ago. (Mention was made of them in chapter 2.) They were those dome-shaped rocks that formed a kind of petrified forest made up of extremely fine layers of calcium carbonate deposited by the billions of individuals living on their surface. Evidently, even in those remote times, primitive beings already found some advantage in grouping together (even if they were not real communities or colonies).

Some examples of primitive colonies can still be found on our planet; these living fossils can provide us with a rather probable representation of what primordial cell colonies were like.

We can see some of them under the microscope. Here are individuals united in gelatin-like masses (*Croococcaceae*), while these bacteria here are organized into long tubes that create filamentous colonies capable of swimming (*Oscillatoriae*). Other spherical cells (*Nostocaceae*) form strings like a pearl necklace. Still others, by uniting, seem to branch out.

Then there are bacteria that form pairs (*Diplococci*), chains (*Streptococci*), and bunches (*Staphylococci*). And larger cells, *phytoflagellates,* also embedded in a gelatinous mass, that are capable of coordinating the individuals composing them. Some are united by filaments, which form a kind of primitive network.

Heliozoans (the protozoa we saw earlier with tests and long filamentous offshoots forming sunbursts) live solitary lives, but when they crowd around a prey, the outer ones suck in the food through the filaments of those that are feeding directly on the prey. An analogy may be the free riders who hook up directly to electric utility grids to withdraw electricity. In the case of the heliozoans, this behavior is protosocial. In a sense, it may be considered a kind of redistribution of wealth.

The force of cooperation

Like a "time machine," the microscope offers us abundant examples of nature's evident and probably very ancient tendency to invent solutions that are not only competitive but also cooperative.

Fossil findings provide us with only some evidence of this, but the course of evolution demonstrates that in the long run cooperative behavior has proven successful, giving rise to more and more complex and organized structures up to the most modern forms of life. In fact, colonies, various kinds of association, and symbiosis represent transitional steps toward those extraordinary pluricellular organizations which gradually led to the birth of plants and animals.

Indeed, everything we have seen in these first three chapters points in this direction: the assemblies of atoms to create molecules, of molecules to create polymers, and of polymers such as DNA and proteins to construct other basic structures of life. The assembly of various structures to build spherules, protocells, and finally bacteria-type prokaryotic cells. The assembly or symbiosis of prokaryotic cells to create eukaryote cells. And finally, the assembly of cells into colonies capable of interacting, with specialized cells that cooperate. The outcome of this long process is colonies of highly organized and specialized cells; that is, pluricellular individuals or, to put it another way, plants, animals, and us.

Shrub-like evolution

It's evening. The sun is preparing another magnificent sunset out at sea. It's already turning the clouds pink and will soon sink behind the horizon. A brisk wind has come up, cooling the air.

On this October evening marking the beginning of a new phase that will lead to pluricellular creatures, we would like to close our journal with a few thoughts on the paths taken by evolution. Let's see if we can introduce some order into the many (often disparate and diverse) things we have seen, and in doing so bring some perspective to the things we are about to see.

It is obvious that the long series of assemblies—from the simplest to the most complex—did not follow a clear-cut course, like a train traveling along one track and stopping at the various stations. The "passengers" who descended from the train (that is, the archaic forms of life that still exist today) actually stepped off branches that were "dead ends" in an endless railway network.

Evolution is by nature "arboreal," that is, it develops in various directions, like a tree. Its branches in turn ramify and develop other branches. Actually, it would be more precise to say that evolution is like a shrub, in that there are so many branches that one can hardly distinguish the trunk.

Sometimes the branches run parallel or cross, diverge or converge. Certain ones dry up, while others again ramify and lead to new possibilities. Evolutionary attempts may fail at one point, only to flourish at another. The same kind of growth may take place quite independently in more than one place at the same time with different outcomes.

For these reasons, it is difficult (especially in the absence of fossil records) to retrace the innumerable branches of the evolutionary shrub and identify the exact chronologies and evolutionary paths followed. And this is why we can only imagine how things must have gone.

It is extremely likely, for example, that the transition from unicellular to pluricellular life occurred innumerable times—at different times and in different places—but that many of these attempts failed without leaving any trace.

This logbook is not the place to sketch a map of the millions of "arms" of evolution's complicated labyrinth. But it is important to keep this labyrinth of "attempts" in mind and to realize that the apparent "jumps" between one living form and another that we observe in nature today (for example, the differences between protozoa that we noticed under the microscope) result from the fact that only certain forms of the entire chain of gradual transformations have survived selection.

It's a little like playing bingo. If we could imagine drawing all the numbers in order (1, 2, 3, 4, etc., up to 90) and placing only the winning numbers on the card in front of us, we would see "jumps," for example, 5, 13, 27, 31, and so on. The same thing happens in nature: there are always missing links between one species and another because the natural filter of selection has rewarded only certain numbers, discarding those that lie in between. The only difference is that in nature, the numbers and the cards are almost infinite.

This is also true of the so-called *theory of punctuated equilibrium,* according to which evolution goes through alternating periods of rest and acceleration, depending on local changes and the rapid spread of certain favorable conditions.

We felt it useful to go over this before continuing on because the number of branches of the evolutionary shrub of life is going to increase at an incredible pace from now on. And of course, we will only be able to look at a few branches, those we feel to be most significant.

With the transition to pluricellular organisms, in fact, evolution is about to undergo strong acceleration. There will be an outburst of forms in all directions. Let's go discover them.

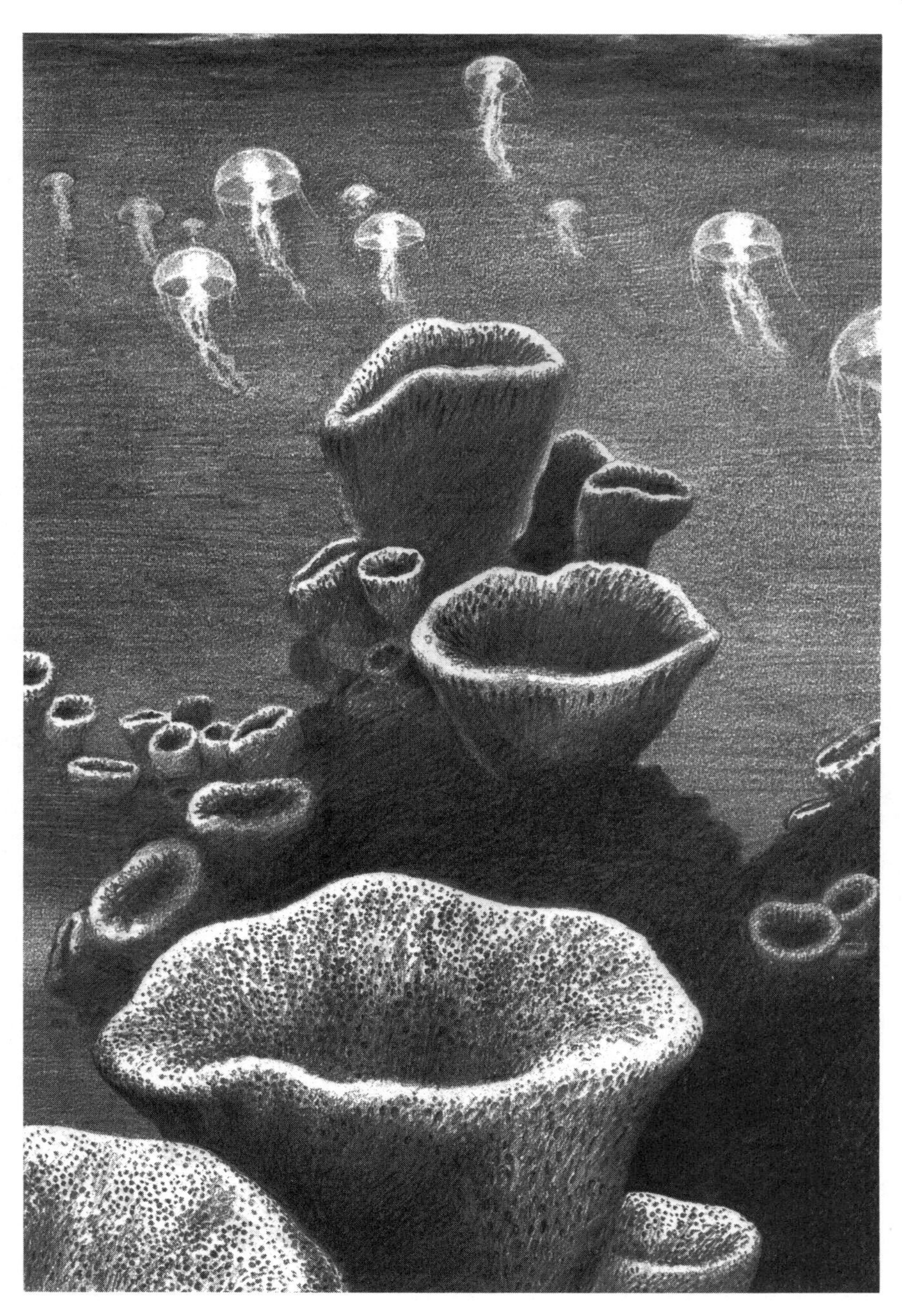

". . . Our attention is drawn by a huge round 'skyscraper' rising from the pale sea floor."

4

The Birth of Pluricellular Organisms

OCTOBER 27TH (700 MILLION YEARS AGO)

Sucked into a tunnel

While we are swimming underwater, on the same scale as protozoa, our attention is drawn by a huge round "sky-scraper" rising from the pale sea floor. Although actually no more than eight inches high, it looks like the Empire State Building from our point of view.

Slowly exploring the outside of this extraordinary structure, we notice that it is full of irregular holes of varying sizes: pores leading to the inside. Upon closer inspection we realize that this is a sponge.

Sponges are probably the most archaic pluricellular organisms on Earth. What we have before us is practically one of nature's first attempts at building a more complex system: a colony of cells with some kind of division of labor. It took a long time to understand that these creatures are not plants but animals.

The shape of this sponge is different from that of the sponges we use in our bathtubs today: this one is vertical, almost tubular.

During our inspection, a particle floating around in the water nearby is suddenly sucked into one of the pores by a strong current. Before we know it, we are being sucked in, too.

Tumbling head over heels, we find ourselves in a dark tunnel dimly lit by light from the inside. The current continues to drag us forward until we reach a kind of huge vertical well. All around us, the walls are lined with flagella moving in the same direction, creating the current that has sucked us in here.

The entire well is lined with flagellate cells. We can see other organic particles and cells being drawn in through the tunnel behind us. All of a sudden it dawns on us: we are meant as food for this sponge.

A dizzying cavity

Swimming upward to try to escape capture by the flagella, we pass by other tunnels through which food particles are being sucked in.

Finally, we reach an area in which there is less turbulence. Not far away is a kind of rocky outcropping with a small pore above it and no flagella: something to hang on to. We grab onto the outcropping and pull ourselves up to peer through the "window."

We can hardly believe our eyes: it's as if we were looking down the elevator shaft of a skyscraper. The cavity plunging down before us is lined with rocky outcroppings similar to the one onto which we are holding: they are the so-called *spicules,* that is, pointed bits of calcium carbonate (or silicon) which form a kind of supporting skeleton for the sponge.

The cavity contains a number of other things: a gelatinous substance (the *mesenchyme,* which acts as an elastic connective tissue) and above all some amoeboid-type cells, that is, tiny sacs that can take on any shape required. In fact, these cells transport the food that has already been "processed" by the flagellate cells and expel it as waste. But they can also serve other purposes: reproduction, lining, or generation of materials needed for construction of the connective tissue or the spicules.

What we are witnessing is the beginning of the kind of malleability and plasticity which will be the basis of future cellular specialization in more organized organisms.

This ability is very limited in sponges: they can do little more than build simple structures made of tunnels and spicules. That is why they represent a dead branch of the evolutionary shrub, a branch that did not give rise to any more complex descendants.

Today, approximately ten thousand varieties of sponges exist. They vary in shape, color, and size (from one thousandth of an inch to two yards), but that is all.

Regenerating from a mash

Despite their simple structure, sponges are capable of some amazing feats. If you take a sponge, crush it and filter it through some fine silk (thus obtaining a mash of isolated cells in the bottom of a water-filled basin), the sponge will soon start to reconstruct.

First, the isolated cells begin to reunite. Then they develop protuberances that link up with others. At this point, the various types of cells take up their original position and work out their old structure: pores, chambers, and so on. After only a week, the sponge has completely regenerated, like a phoenix from its ashes.

The surprising thing is that—even if we don't realize it—the same kind of marvel takes place in our bodies: if you make a mash of young epithelial tissue, that is, the tissue which lies on top of the dermis, the underlying layer of skin, the isolated cells will link up anew and slowly reconstruct the tissue. This is how the fine layers of skin used for grafts are created: by regenerating epithelial cells onto an appropriate support.

Reconstruction is even more amazing in the liver: if part of the liver is surgically removed, the remaining cells reconstruct perfectly the complicated original structure (bile ducts, the lobes, blood vessels, etc.), recovering its many original physiological functions.

Hermaphrodites

We are still climbing the well. Other particles are gently wafting around. It has been estimated that a four-inch sponge filters twenty-five gallons of water per day to obtain nourishment for its cells. The channels and tunnels function like "inverse" intestinal villi: they increase the internal surface of the walls and make absorption more effective.

The oxygen that the cells need to breathe is drawn in with the water. Waste products are then expelled, together with gametes and larvae. Sexual reproduction is, in fact, one of the ways in which sponges reproduce, even though they have no particular sexual zones: the gametes (reproductive cells) can develop anywhere. Sponges are basically hermaphrodites. The male sex cells move around the structure and enter into the chambers where they meet the female cells (*oocytes*). Fertilization then takes place, producing larvae.

One larva is just about to leave the top of the well. Looking like a small colony of floating flagellates, it slips over the edge and floats away gently on the current.

We are now approaching the summit of the well ourselves. It terminates in a circular opening: the so-called *osculum*, or mouth, which has a system for closure, or rather, obstruction that is regulated by special cells along its edge.

Perched on the rim, we can feel mild contractions running through the sponge's body. The sponge has no nervous system or special system of communication. Its most direct means of communication are its canals, which penetrate everywhere. The osculum, however, has a primordial sensitivity to touch. Tapping lightly on its edge with our small geological hammer, we can feel the osculum close slightly. The stimuli are then transmitted downward from one cell to the other.

Without dedicating further time to scientific investigation, we make for the open seas and for freedom once again, only too happy to leave this nightmare of tunnels and canals behind. We certainly would not recommend the experience to anyone suffering from claustrophobia.

Behind us, a single sponge cell drifts through the partially closed osculum. This cell has separated from the sponge. Sooner or later it will attach to the sea floor and give rise to another identical sponge.

This is the other way in which sponges reproduce: by *gemmation,* that is, the division of a normal cell. This dual system is obviously very advantageous: under adverse environmental conditions, such as unusually low temperatures or serious drought (for sponges that live in fresh water), gemmated cells are more resistant and can merely wait for more favorable conditions before developing.

The Peter Principle

Gazing back on our strange "skyscraper," we can't help thinking of the well-known Peter Principle, which states that in any kind of business or administration, employees are promoted until they reach their level of incompetence and that is where they remain for the rest of their lives. For example, good accountants may be promoted to head of the department, meaning that they will leave a job that they know how to do well for one that they may not know how to do at all. If they are good at the job of head of the department, they may be promoted to an executive position, and so on. According to the Peter Principle, then, everyone moves up until they arrive at the position in which their incompetence is manifest, which is where their career ends.

The sponge reached its own "level of incompetence" in a very short time (only 700 million years). It has not had and will never have an "evolutionary career" beyond that because it can do no better than it is doing at present. It will never pass on to subsequent stages; this is where its development stops.

Certain sponges made attempts in the right direction, but they lacked the requirements needed for further development. For example, some took on radiate forms, but to no avail. The radiate form is, in fact, one of the most important steps (necessary but not sufficient) for evolution toward more complex organisms. It involves endowing the body with a geometry that makes it not just an agglomeration of cells, but an organized system. (Jellyfish are an example of an organism that managed to take this step successfully.)

NOVEMBER 5TH (600 MILLION YEARS AGO)

The birth of an island

Back in our tent on the promontory, we try to reorder our notes and our thoughts.

We have spent the last few days measuring the oxygen concentration in the atmosphere: it is now over 7 percent; ozone is also beginning to form.

Out at sea, about twenty-five miles to the northwest, a column of black smoke rises into the air. We first noticed it this morning when we woke up. An island seems to be forming. It can be seen quite distinctly through our binoculars.

The eruptions of an underwater volcano have accumulated so much lava that the cone has finally emerged above the surface of the water. The crater, just over half a mile in diameter, is now spewing incandescent rocks and thick black smoke into the atmosphere. When the eruption ends, this volcano will form a new island. Two others have recently been created in the same way just a few miles away.

Luckily, the wind is blowing in the other direction, carrying the fumes away from us. But we can still feel the tremors underfoot every now and then.

It's a beautiful day and the sea is calm after last night's storm. Down on the beach, an oddly shaped object that is being rolled back and forth by the waves catches our eye.

It can't be a stone; it seems to be changing shape as it rolls along. It looks soft. We've never seen anything like it.

Scrambling down to the beach, we cautiously approach the small gelatinous mass that sparkles in the sun as it is washed to and fro by the waves.

The biological umbrella

It is a jellyfish—a dead jellyfish, but a whole one. It is about as big as a small soup dish with its long appendages protruding from below the umbrella half hidden in the sand.

We run our fingers over it delicately: it's almost a pleasure; it feels like rubber, smooth rubber. The translucent white color is tinged with pink.

Another wave rolls the creature gently forward, revealing the tentacles which extend from the so-called *manubrium*. The filaments hang from the edge of the umbrella: they contain the sense organs and the famous stinging cells that can kill small prey.

Under the umbrella, we can clearly see the characteristic radial structure of this animal (because it is an animal, even if early eighteenth-century naturalists had some doubts about classifying it as such). Unlike sponges, radiates have a very well-organized structure, and above all some very advanced cell specialization.

So this small biological relic is the first marine animal to come onto dry land. Admittedly, it has done so in a rather unorthodox way. But still, in this totally desert environment, the waves have brought to light a surprising testimonial of what is taking place under the sea's surface.

Although it is still rather stiff, thanks to its gelatin-like filling, the jellyfish will soon decompose under the sun's rays.

As for us, it is time to dive back into the sea to discover the new forms of life that evolution is generating.

The first men-of-war

This time we don't have to scale down to cell size. On the contrary, it would be dangerous to do so because the seas are now full of terrible predators such as jellyfish and their various kin. So we reduce to the size of a sardine: just small enough to see the details, but just large enough to avoid being devoured immediately.

Swimming close to the water's surface, we can see multitudes

Opposite page: Right: A Portuguese man-of-war. Bottom left: A modern jellyfish. Fossil remains of jellyfish (above) dating back 650 million years have been found in the Ediacara deposits in Australia.

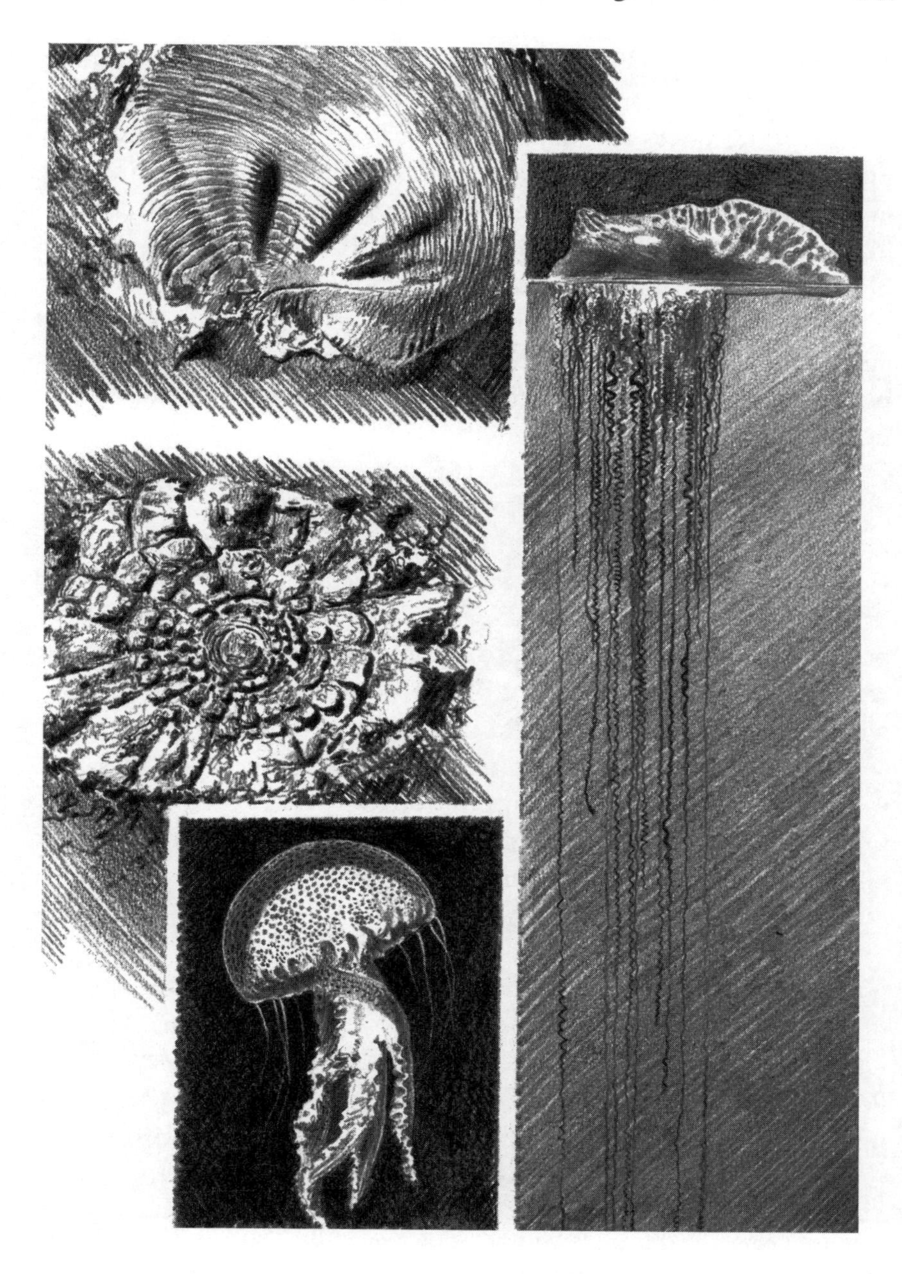

of increasingly diversified protozoa. A huge cactus-shaped sponge is attached to the sea floor, its "arms" uplifted like a preacher's. In spite of differences in shape, all sponges function on the same principle: they suck in and filter great quantities of water, absorbing anything that happens to be nutritious into their system of villi-like canals. Basically, they are like small bits of intestinal tract planted onto the sea floor waiting for food to come their way.

Suddenly a strange jellyfish with long trailing filaments appears in the distance. The tiny creature is floating on the surface, but its appendages dangle in disorder like a cascade of branches from a flower box.

Actually, this is not a jellyfish, but a "Portuguese man-of-war" (*Physalia*), so named for the fact that it floats on the surface. The buoyant colony is made up of very specialized individuals: some cells secrete gas which keeps the colony afloat; others have a digestive function; others reproduce; still others produce a poison with which to attack and defend. This is a more complex version of the story of the blind man and the cripple, with entire colonies of individuals cooperating.

Scientists believe that animals of this kind (the Portuguese man-of-war is only one example) derive from medusoid (that is, jelly-like) beings or from fragments of colonies that broke off and then evolved independently.

The hydra's tentacles

We continue our exploration, constantly on the lookout for predators. They could appear at any time and, given our dimensions, would not be a very welcome sight.

Opposite page: Hydra are like upside-down jellyfish. They attach to the sea floor with their tentacles reaching upward. The sequence illustrates the positions they may adopt when moving.

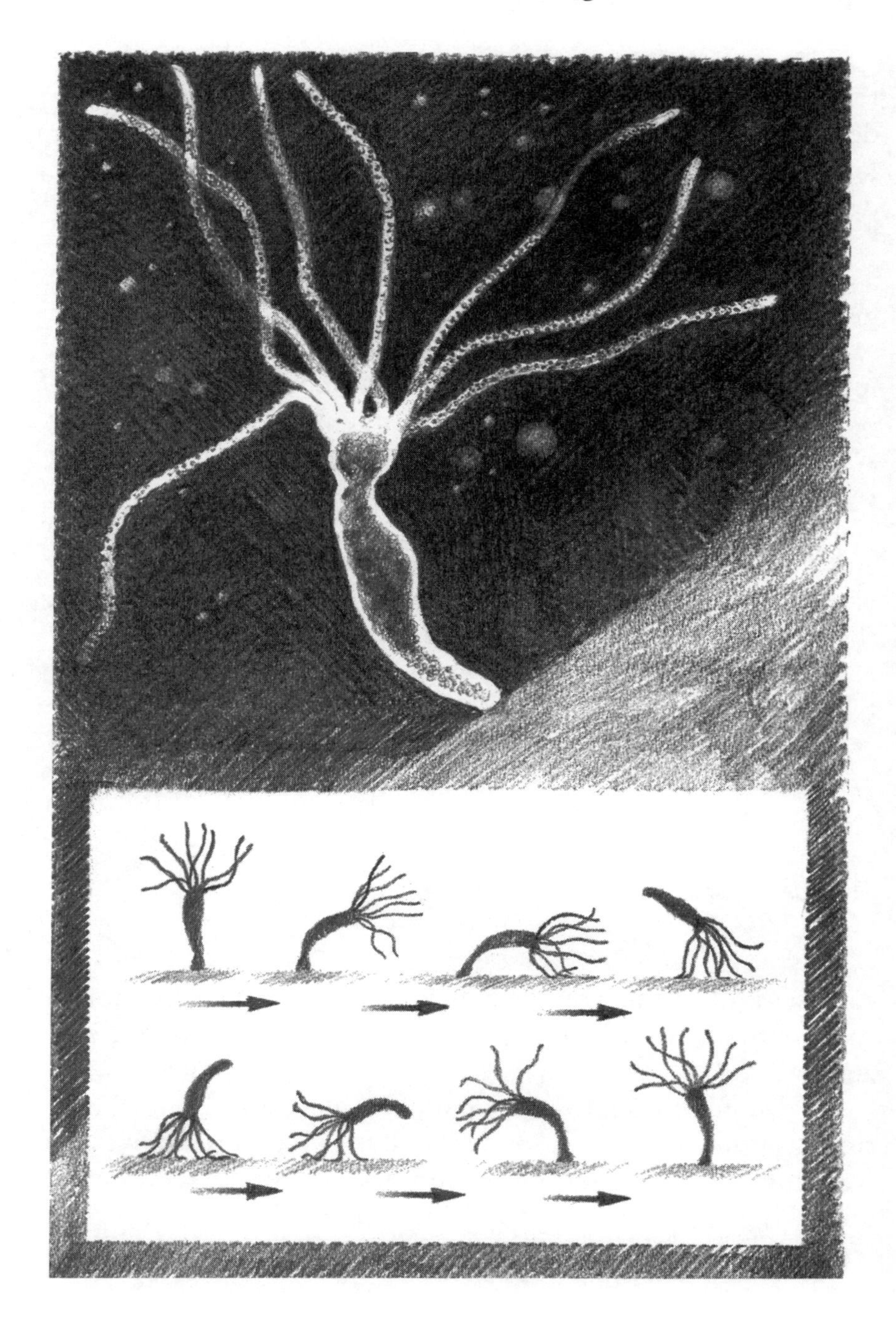

The rays of the sun penetrate softly through the limpid waters. There is no turbulence, no sand: visibility is excellent. Suddenly we see another unusual form at the bottom of some shallow waters: it looks like a flower with a crown-shaped corolla. We decide to swim over to have a better look.

It's not a flower at all, it's an animal. The corolla is actually a group of tiny tentacles—eight to be exact. At the center of the group is a minute cavity: the mouth. This is a hydra. The hydra resembles an upside-down jellyfish attached to the sea floor with its tentacles reaching upward. But its movements are well coordinated: this is a very sophisticated organism, with some basic systems that will be developed in superior organisms. It is on the right road to evolution.

The hydra has sensitive cells (especially around its mouth) and a primitive muscular system. But the most exciting novelty is the presence of nerve cells, localized in the gelatinous cavity, that constitute a primordial neural network.

Of course, it has neither a brain nor ganglia. But the sensitive surface cells are linked to the underlying nerve cells, and this creates a primitive sensory-nervous-locomotive system. What exactly does that mean?

Armed with a pin

A closer look at the hydra may help us explain. If we had something with us with which to jab it, we might be able to see its reaction.

We've gone back to pick up a pin which, given our dimensions, is now the size of a banderilla. By poking the sensitive area we should be able to set off a circuit: the cells of this area should pass a signal on to the underlying nerve cells, triggering the transmission of impulses throughout the entire system (a very elementary system, to be sure, but one that already has axons, dendrites, and synapses, that is, the basic components of the future central nervous system). When running through the system, this signal should stimulate certain muscles and make them move. This three-tier combination is the sensory-nervous-locomotive sequence which lies at the root of all elementary behavior.

Clutching the pin like a javelin, we approach the hydra trying to keep out of reach of its streaming tentacles. After having identified the area with the highest concentration of sensitive cells—those around the mouth—we take aim and hurl the pin. The force of the blow is blunted by the water, but the pin hits its mark nevertheless. The hydra contracts; the reaction is probably more one of surprise than of pain. The tentacles retract into a defensive position and the pin falls to the sea floor.

We rapidly withdraw to a distance from which we can observe the hydra's behavior. After a few seconds, the hydra starts to dilate again. Then it begins to bend as if it wants to move away.

Before our astonished eyes, it executes a curious backbend, taking on a horseshoe shape. Then it straightens up again, but with its tentacles on the bottom this time: it is standing on its head. It immediately bends over again and straightens up to find itself in a vertical position once more. Then it repeats the whole procedure. Basically, the hydra has just completed a series of handsprings. And in this way, it moves off.

Surprised, we reflect on its reaction: a simple pinprick was enough to set off this very complex behavior.

We decide to stay in the vicinity to see what else will happen. The hydra seems to have calmed down and has once again taken up its normal position. Its tentacles are relaxed and float gently in the water.

But now a larva (a jellyfish perhaps? or a hydrozoan?) is drifting toward it. The tiny colony of cells is looking for a place to anchor to undertake further development. It has no idea of what is in store for it.

As soon as the larva gets within range, the hydra's tentacles start to swirl around and the eddy of water created by that motion deposits the larva softly near its mouth. The muscles of the hydra contract and the larva is swallowed. Inside, enzymes enter into action to digest the prey.

But let's leave this strange animal and look for a jellyfish. It shouldn't be too difficult to find one, as these are the waters that will one day become the famous Australian Ediacara deposits.

An encounter with a giant jellyfish

While crossing a sunny shoal, we saw a number of other hydras of different sizes, all with their tentacles turned upward. Food is beginning to be plentiful in these seas: there are not only tiny protozoa, but also appetizing larvae and various types of other (still very simple) animals.

But we still haven't seen any jellyfish like the one we saw on the beach. And yet jellyfish are not solitary animals; if there is one around, there should be more.

We change direction and make for a shadier area. Indeed, something seems to be moving ahead. It becomes more distinct as we approach: it is a giant jellyfish. We have no idea of its actual size, but from our perspective, its umbrella is as large as the dome of a cathedral. It's quite a spectacle.

Filaments hang from the perimeter of the umbrella-like tassles from the shade of an art nouveau lamp. The tentacles extend from the central manubrium. The jellyfish's movement is extremely graceful. In the water, it looks almost transparent. Now and then its umbrella pulses slightly as if it were flying: the gelatinous mass contracts while the filaments and the tentacles follow suit in a wave-like movement. It is a beautiful sight, the most beautiful that we've seen since we've landed on this planet.

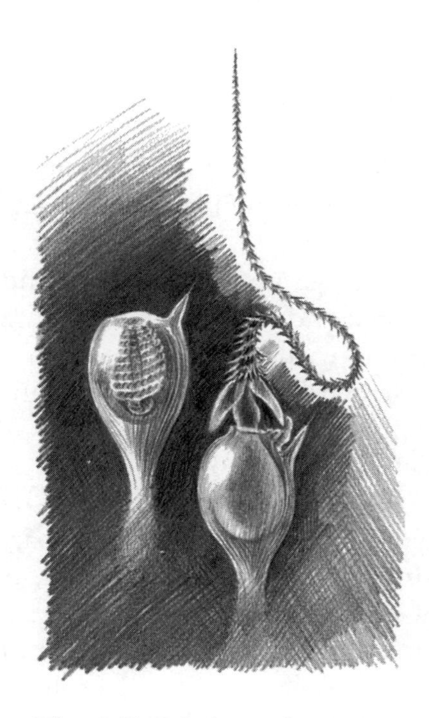

The attack mechanism of a jellyfish. On the left: The poisonous filaments are rolled up inside a closed capsule. Upon stimulation, the capsule shoots its "harpoon" (right).

The jellyfish is no longer contracting. It is simply letting itself drift with the water. We can't resist the temptation to inspect it more closely.

Compared to the hydra, the jellyfish looks like a glove that has been turned inside out. But despite considerable differences, the two species are basically the same animal. Like the hydra, the jellyfish has sensorial cells, a neural network, and muscle cells and lacks organs: it has no respiratory, circulatory, or excretory systems (that is, it has no equivalents to our lungs, heart, intestines, brains, etc.). It nevertheless breathes—through a membrane—taking in oxygen and giving off carbon dioxide, like human beings. Like the hydra, it has "testicles" and "ovaries," or, more precisely, two sacs, simple protuberances that contain the male and female reproductive cells.

The harpoon attack

The jellyfish is now moving very slowly. We decide to position ourselves above it to get out of the range of its filaments and tentacles.

From above, it looks like a huge blown-up sheet, swollen and translucent. We drop down to touch it and slide down the dome almost to its edge, where the frightening filaments with which it captures its prey are located. At close hand, we can see various rows of them, like missile batteries.

The jellyfish uses a hunting method similar to that of whalers. Indeed, whale hunters use guns to fire harpoons that are attached to strong ropes. While flying towards its target, the harpoon unravels the rope. Sometimes the tip of the harpoon is coated in a paralyzing liquid that numbs the victim.

Jellyfish (like hydras and other medusoids) have a number of stinging (*cnidocyst*) cells that contain a special fluid and a rolled-up filament. The surface of each of these cells has a kind of button which, when touched, fires the filament.

There are a number of ways of capturing a prey: the filaments can either penetrate it (like a harpoon) or wrap around it (like the *bolas* used by South American Indians). They then inject it with their venom, a hypnotoxin which induces paralysis. Some jellyfish use gluey substances that turn the filaments into a kind of sticky flypaper from which the victims cannot free themselves.

Of particular interest is that the button is tripped automatically by the presence of certain chemicals. In other words, the prey need not necessarily touch the button; contact with certain molecules (that raise the osmotic pressure inside the cell) emitted by the victim are enough to trigger an automatic attack. The reaction is similar to that of an animal pricking up its ears when it smells an odor: the jellyfish automatically fires its filaments when it senses the presence of a prey in the vicinity.

In fact, a harpoon has just gone off.

The filament wraps around a strange transparent creature that was passing by. The attack is so sudden that we didn't even have time to notice the prey which is now being dragged under the umbrella toward the jellyfish's mouth. It is difficult to make out what is happening from our point of view, but it doesn't take much imagination to see the victim being converted into proteins, fats, and sugars by the jellyfish.

A part of these substances will serve to replace the harpoon. In fact, once the harpoon has been shot, it cannot be rewound; the same goes for the cell that hosted it. It will be digested and replaced by another cell that migrates to the designated area from the derma. In the meantime, however, many other cells will be ready for action.

We decide to swim toward shore while the jellyfish is still occupied with its meal in order to keep from becoming dessert.

The richness of sexuality

Coming to a new area, we are met with a very unusual sight: the entire sea floor is covered by a forest of hydras. This is the first time since we set out on our journey that we have seen a landscape so rich in life, shapes, movement.

The forest also contains other colorful flower-like creatures whose petals float gently in the water. These *actiniae,* or sea anemones, will be the first "antozoa" (that is, animals in the shape of flowers) to populate these seas together with corals. They are yet another variation on hydras or jellyfish and have similar characteristics.

As we gaze in wonder, some tiny medusae, no more than a few millimeters long, rise from the bottom. They have, surprisingly enough, been ejected from the protuberance of a hydra. It is as if the hydra has given birth to a host of tiny medusae, both male and female. By freeing their gametes (i.e., sex cells), which will unite with others, the medusae will, in turn, give birth to new hydras. In this case the glove is turned inside out twice.

Some planulae also drift by. They represent intermediates in the reproduction of certain jellyfish. They are, in fact, swimming ciliated larvae which, at some point, attach to the bottom

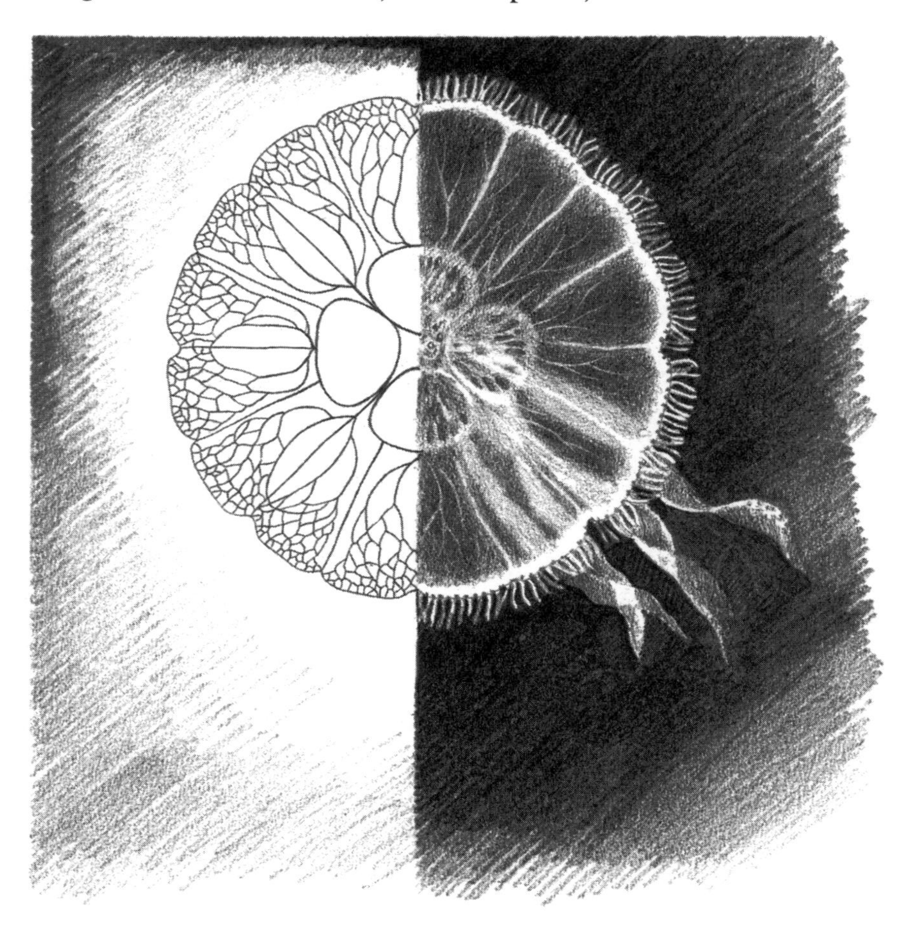

A jellyfish, *Aurelia aurita*. The lefthand side of the illustration shows the nervous system reflecting the animal's radial symmetry.

and develop into what look like tiny octopi. Then they grow into organisms that resemble a stack of jagged-edged plates and finally detach from one another to become adult jellyfish.

So, new and extraordinary methods of reproduction are developing in these seas, with a series of innovations and variations that foretell the future range of sexuality in living creatures.

The actiniae are also reproducing now. Some have external pockets that resemble primitive marsupials in which the young grow. Others have different systems of reproduction: they break the small supporting disk which keeps them anchored to the sea floor and move on, leaving pieces of the disk to reconstruct complete adult individuals (a kind of duplication which brings to mind the famous scene from "The Sorcerer's Apprentice" in Walt Disney's *Fantasia,* in which the pieces of broom regenerate new ones).

Toward a nervous system

Our impression upon reaching shore is that the seas are now teeming with life. True, some of the major protagonists are still missing, but what we saw in this long swim has been quite thrilling: not only is sexuality abounding in the most diverse forms, not only are creatures hunting with harpoons and poisons, but the cells that will mark a turning point in evolution—nerve cells—have appeared in the delicate gelatin of the hydra, the jellyfish, and the actiniae.

For the moment, they are no more than simple cells linking the skin to the muscles, but networks are already being shaped to coordinate overall movement.

The next decisive step will be the appearance of neural ganglia, that is, those "tangles" of nerve cells that multiply an animal's capacity for coordination. And that will probably lead to the most important event in the history of life: the advent of the brain.

“The light dims as we descend and semidarkness closes in around us. . . . But life can now exist without the sun. . . . We turn on our flashlights and the cones of light zigzag down the cliff as we descend.”

5

In the Warm Seas of the Cambrian

NOVEMBER 8TH TO 15TH (570–505 MILLION YEARS AGO)

Italy at the South Pole

Over two hundred million years have gone by and the seas are bursting with life. These are the seas of the Cambrian period, a crucially important and classic period for paleontologists, given its abundance of fossil remains (in fact, the Cambrian takes its name from Cambria, a region in Wales, where the first fossil deposits were found).

Some of the most fundamental transitions in evolution will be taking place in this period, with the emergence of the great lines that are to lead to crustaceans, fish, insects, amphibians, and all the forms of life that will later populate the continents. The continents themselves are quite different from the land masses we see in the atlas today: they are upside down and largely still welded together.

Pangea ("All Lands"), the vast supercontinent comprising all existing land, has split and the resulting continents have drifted to places which we find unusual: South American and Africa, still together, are at the South Pole. Italy, along with Greece and Yugoslavia, is also in the same vicinity, close to the Antarctic Circle. Not far away is Antarctica joined to India and Australia. This huge land mass is known as Gondwana.

The rest of Europe is off in a completely different direction, between the tropics and the equator, close to North America; Asia, split in two, lies almost at the North Pole together with Borneo. This planet is unrecognizable for anyone looking at a modern map of the world.

But the Cambrian seas are warm and comfortable and it is a pleasure to swim in them. A pleasure which is heightened by the intense contrast between the terrestrial environment—still a barren volcanic desert—and the marine world, abounding with life.

A lot of fingerprints

The essential difference for paleontologists is that evolution has begun to leave "fingerprints." Life is starting to develop shells and casings of various sorts: structures that fossilize. After being "soft" for billions of years, life is now starting to construct the first scaffoldings for support and defense. These structures, which are destined to become skeletons and exoskeletons, will also leave more and increasingly clear traces in the sediments. This is why the fossils from the Cambrian period represent the first illustrated book of life, the first photo reportage of evolution, after the rare scraps found from the preceding eras.

Many forms still do not have rigid structures: in the modern era, over 50 percent of the large groupings (*phyla*) of life still have only "soft" parts and this proportion was probably even higher in the Cambrian period. Therefore, a large number of animals have disappeared without leaving any trace. But those that we can observe provide a very significant picture of this phase in the history of life.

All we can do is dive in to take a look.

A submerged plateau

Stretching out in front of us in the shallow water is a kind of underwater plain or, rather, plateau, as there is a steep embankment slightly further on that drops off sharply into the depths.

". . . Strange colonies of tiny orange-colored octopi shaped like birds' feathers sway gently as we pass."

The sea floor is covered with fine sediment, basically silt produced by erosion on land. A couple of jellyfish float by graciously, letting themselves be carried by the current except for the occasional pulsing of their umbrellas. A whole school of tiny medusae drifts by, resembling a ballet company in tutus.

Suddenly we reach a dark line cutting across the sea bottom: the edge of the underwater cliff. We swim up to it and peer down into the dark depths. It gives us a curious sensation, like looking down from a very high bridge.

Below, we can barely make out a tiny form as it darts behind an outcropping. Farther down still, just within our visual range, is a large, rather strange-looking creature that glides out of sight with a wave-like movement along its edges. Diving off the bridge, we let ourselves sink slowly, fanning our arms every now and again to regulate our fall.

The cliff is extremely steep. Its surface is dotted with sponges and various kinds of *pennatulae* (strange colonies of tiny orange-colored octopi shaped like birds' feathers) that sway gently as we pass.

As we descend the light dims and semidarkness closes in around us. The water also becomes much colder. But life can now exist without the sun: the seas are rich in oxygen, and breathing through the skin is more efficient, allowing for, among other things, a more active metabolism and greater mobility.

This is a dark and unknown world inhabited by primordial forms. Even though we know that large predators have still not appeared, entering it is nevertheless somewhat disquieting, as the strange creature about a yard long that we saw disappear into the depths just before our dive is still in the back of our minds.

We turn on our flashlights and the cones of light zigzag down the cliffs as we descend.

At the bottom of the cliff

Every outcropping is covered with traces of the silt that has floated down and accumulated. Its extremely fine texture is highly suited to fossilization, making it possible to preserve even the most delicate structures, such as the soft parts of organisms. In fact, a large number of extraordinarily intact fossils (in which even the hairs of certain worms are visible) have been found in this place (known among paleontologists as the "Burgess argilloscyst formation" in Canada).

Our flashlights are now pointed at a curious object: it looks like a slice of pineapple with a hole in the middle. This is a *Peytoia,* a swimming coelenterate, related to the jellyfish. Under

A strange object appears: it looks like a slice of pineapple with a hole in the middle. It is a *Peytoia,* a swimming coelenterate.

the spotlight in this dark setting, it seems to float in space like one of those wheel-shaped space stations that are sometimes depicted in science-fiction movies.

The *Peytoia* drops out of sight as we continue on our way down. According to our depth counter we have already descended 450 feet.

Strange sculptures now come into sight: we are approaching the bottom. The cliff gives way to a huge plain. This could be the setting for a Western movie with cactus-like sponges rising out of the muddy desert: some candelabra-shaped, others resembling vases with tufts at the top, others hairy cigars. What a strange underwater garden; in the darkness, its bizarre and faded forms take on unsettling contours.

Finally we hit bottom, raising a murky cloud which slowly diffuses through the water. Despite the total darkness and cold, life seems to have adapted well to these conditions.

Strange armored creatures

We set out to survey the plain, paddling through the sponges. The silty bottom is speckled with small holes: the "lairs" of primitive worms that live buried under the mud. A large variety of these worms still exist today. They are *polychaetes* ("many-bristled"), basically a marine version of the earthworm.

Hidden in their burrows, they take advantage of food in the vicinity by filtering the sand for organic compounds. They must take care, however, not to fall into the trap of the *Ottoia,* a voracious predator with a retractile trunk covered in spines that juts out of the mud.

Fossils reveal that the *Ottoia* attacked everything in its range: even the remains of the conic shells of primitive mollusks have been found in its stomach. As its teeth were not strong enough to break the shell, the *Ottoia* simply swallowed the organism whole and, after having digested the mollusk, expelled the shell. Remains of individuals of its own species have also been found in its stomach, a sign that the *Ottoia* practiced cannibalism.

Before us stretches a small trough, like the track made by a wheel. Following it for a couple of yards, we come upon a strange creature resembling a lady's hat decorated with feathers and pins. This is a *Wiwaxia,* possibly an ancestor of the snail. Right now, as it crawls along the sea floor, it is probably gathering food with its raspy tongue whose countless tiny dentelles rake through the sediments. We watch it in amazement as it moves away: a tiny tank pulled by an invisible wire.

Another armored creature that immediately catches our attention is a primitive crustacean with two plates that make it look like a slipper. A tail composed of horny rings extends from the back, while two small "antennae" project from the front.

The scientific name for this creature is *Canadapsis.* An inhabitant of the sea floor, its armor is proof of the fact that the

Wiwaxia

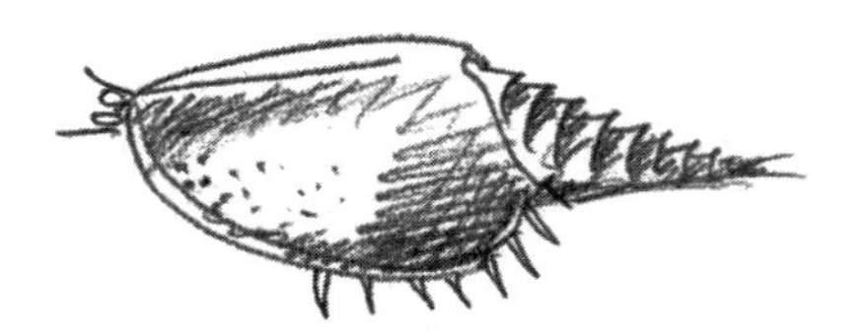

Canadapsis

seas are now populated by dreadful predators, which call for plates of armor and spines for defense.

One creature which seems to have no fear is the *Aysheaia,* a peculiar walking worm that, as one researcher pointed out, looks like a cross between a centipede and a Michelin man. It marches along on short, stocky legs, without any protection or armor for its long body which tends to thicken in the middle. This model will be very fortunate in the conquest of dry land; in fact, this animal is probably among the ancestors of the first forms of life that walked the continents. But many millions of years (and numerous adaptations) will have to pass before the undertaking is a success.

The most striking thing about the *Aysheaia* (making it important from an evolutionary point of view) is that it possesses some of the characteristics of a worm, some of a spider, and some of an insect. In fact, various parts of its body (eyes, trachea, heart, intestine, excretory and reproductive apparatuses) have components that are typical of different phyla, almost as if this animal occupied an intermediate position.

Following a trilobite

Out of the darkness—possibly attracted by the light from our flashlights—advances a tiny flat submarine: a trilobite. Its shape is well known from the many trilobite fossils that have been found (and are sometimes even used as paperweights).

The *Aysheaia*. Top: the fossil imprint and (center) a reconstruction of the animal. Bottom: One of its modern descendants of the genus *Peripatus* which lives among the decaying leaves of the tropical rain forest.

From a biological point of view, the trilobite is a very well-designed machine. Indeed, its many shapes and forms are all very successful in the Cambrian seas. This specimen is approximately two inches long, but trilobites can reach lengths of up to two and a half feet. Some swim, others live on the cliffs, still others move below the sand like underwater moles. They dominate the tidal waters, as crabs do today.

This trilobite has changed direction, possibly disturbed by our presence. As it turns, we can clearly distinguish the antennae on its head and tail. They are sensors capable of perceiving everything that goes on in the environment. The head looks like the rounded hood of a car with two old-fashioned headlights: the eyes.

Although we observe the trilobite for a while as it zigzags across the sea bottom like a hovercraft, we fail to get a look at the most important part of its body—-its belly. This is where its secret weapons are hidden: eight pairs of legs (for movement and for capturing prey) and a series of tiny "combs" on each leg (the so-called *filamentous gills*) for respiration: almost a miniature version of the Indian goddess Kali, with limbs allowing it to walk, swim, grasp, and sense in an extremely coordinated way.

The trilobite suddenly dives toward the bottom; it must have spied a prey. A small cloud of mud rises and the trilobite reemerges clasping something between its forelegs: a tiny worm which it soon tears to pieces and eats.

This is the first act of predation in the modern sense of the word that we have witnessed. And it is to be the precursor of many more.

Opposite page: Trilobites were a very successful group that survived in the primeval oceans for over 300 million years. While maintaining the same basic structure, they developed diverse shapes ranging in length from a fraction of an inch to two and a half feet. Trilobites went into extinction at the end of the Permian period, approximately 250 million years ago.

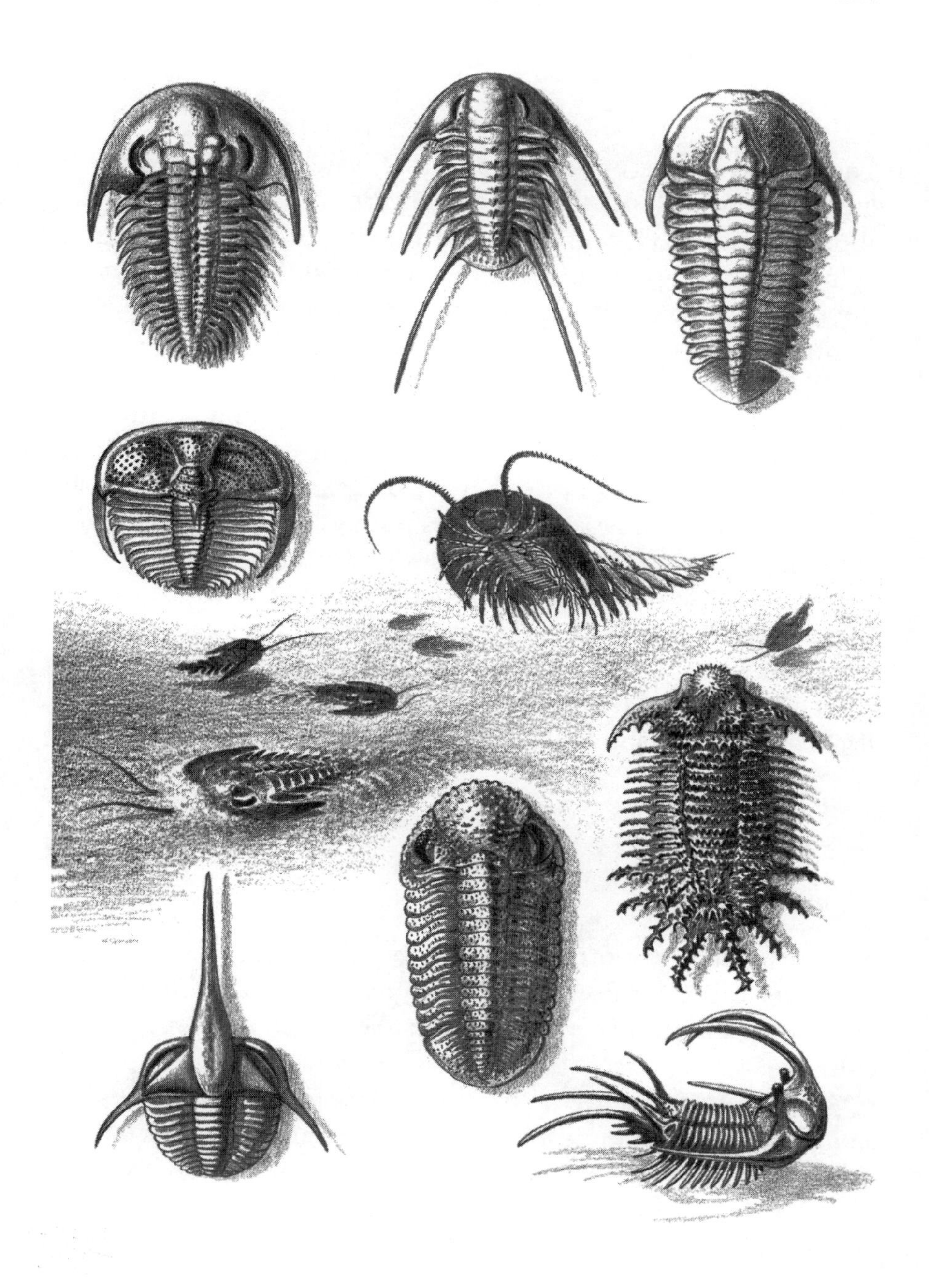

The hunted predator

There are trilobites everywhere and we watch them circle under the light of our flashlights. The sea bottom now offers them an abundance of food. Suddenly, like a stroke of lightning, a huge light-colored shape enters the cone of light and, targeting one of the trilobites, seizes it with two huge dentate legs that project from its jaw.

We can see the large, odd-looking animal more clearly now as it stops to devour its prey which is still thrashing around. Its large armored head is set with two button eyes. It has a long streamlined and rigid body that tapers like that of a cuttlefish. On its underside, a series of paddles resembling the oars of a galley ripple in the water, like a flag in the wind.

Its scientific name is *Anomalocaris* and it is the terror of these seas. This individual is approximately three feet long. It may have been the creature we saw when we were descending.

The *Anomalocaris* slowly swims away. We follow it with our flashlights, but it suddenly disappears into the dark.

The mystery of the Hallucigenia

Surveying the sea bottom we discover other new forms of life: some are calice- or flower-shaped like sea lilies (distant relatives of urchins and starfish). Others look like halved *scolopendra* with long barbs.

As we approach the cliffs again, our attention is drawn by a cloud of silt spreading through the water. An enormous stream of mud is descending the rock wall. Advancing at an amazing speed, the black mud slide is tearing away everything in its path.

A small creature has just been swept up by it. This animal is

"The *Anomalocaris* is the terror of these seas. Approximately three feet long, it has a series of paddles on its underside that move in sequence like the oars of a Roman galley. . . ."

so unusual that when it was first discovered in fossil form its incredible appearance earned it the name *Hallucigenia*.

In paleontological illustrations it is generally depicted as a sausage resting on seven pairs of divergent spines or legs. Its back has a row of tiny proboscises thought to serve for capturing food.

Zoologists long pondered the origins and lineage of this unique creature, which seems totally unrelated to preceding and

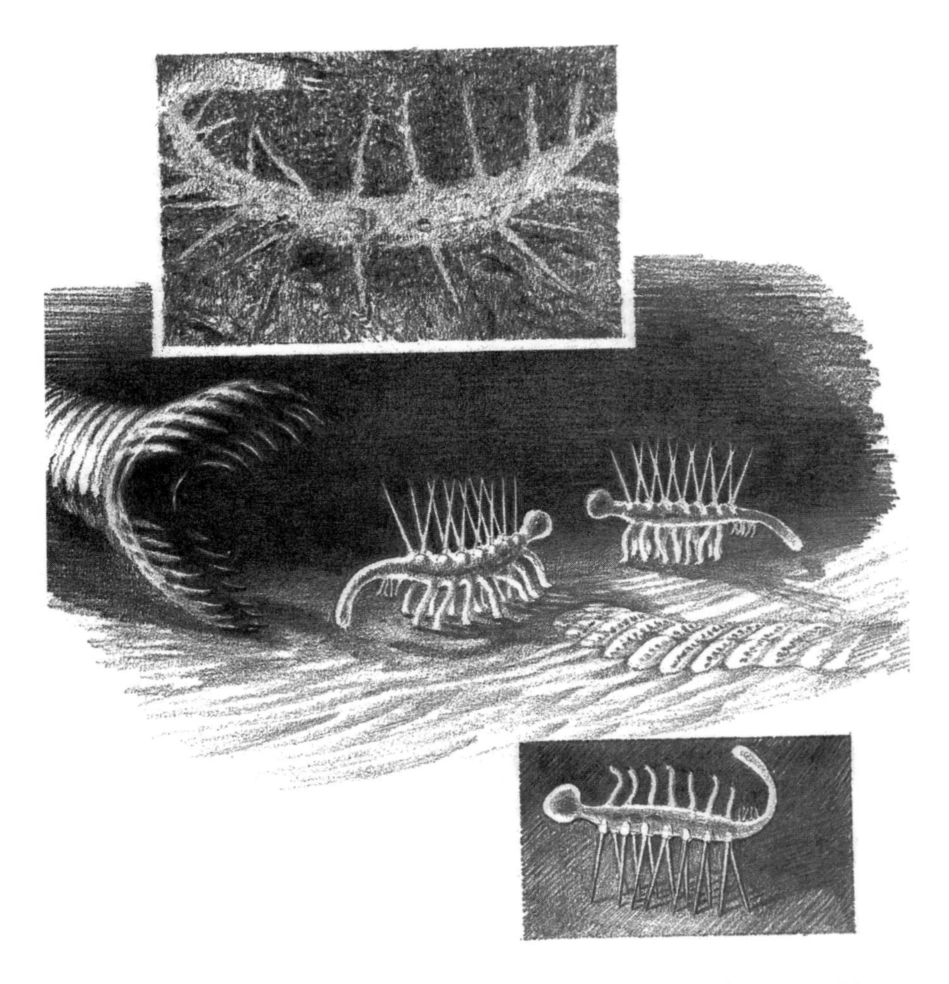

Top: The fossil imprint of the *Hallucigenia,* whose name is eloquent with regard to its appearance. In the past, this small animal was depicted as in the drawing at the bottom. But someone recently decided to turn it upside down and it is now thought to have lived and moved as shown at center: on sturdy legs with its spines projected upward for defense.

"We now see a startling creature slithering through the mud. It looks like something out of a horror movie: it is an *Opabinia*."

subsequent fauna. How could it move in the mud on those fourteen sticklike legs? And how could its proboscises reach the food on the bottom?

Only recently did someone propose a simple solution to the problem: by turning the animal upside down. That way, the proboscises are actually legs and the rigid "legs" are spines for defense (rather like a porcupine). This proposal seems to make more sense. In fact, many other animals in this environment have armor or spines.

The problem may have been generated by the fossil findings, which only reveal one row of proboscises. But some scientists now claim that the other row is simply hidden in the rock below.

Unfortunately we have missed our chance to clear up this minor mystery: the *Hallucigenia* has been buried in the mud.

A science-fiction monster

The mud slide has already covered an extensive area close to the cliff and is continuing to expand. The animals that die and are covered by this mud will be in ideal conditions for fossilization. Suddenly, the implications of this statement dawn on us and we move away as quickly as possible, not too keen on becoming fossils ourselves.

A startling creature now slithers through the mud nearby. It looks like something out of a horror movie: its head is adorned with five eyes gathered into a bunch; it body is segmented and ends in two vertical rudders. But particularly interesting is the long tentacle extending from its head which terminates in a trap-like fang-filled mouth. No science-fiction author has ever come up with anything quite like this.

The *Opabinia* (that is its name) may use this prehensile organ to capture prey by projecting it forward with the precision of a chameleon's tongue, or it may use it to dig into underground burrows. Unfortunately, we are not given the chance to see it in action.

Our investigation is almost over. But the last thing we see at the foot of the cliff is a creature that may have played a fundamental role in the evolution of life toward vertebrates (and therefore human beings): the *Pikaia*.

A fundamental invention

This animal looks like a long thin silver leaf. It is lithe and quick and possesses a unique characteristic: a notochord running down its back—one of evolution's most important inventions. The notochord is, in fact, the precursor of the spinal cord. It is a fine,

Top: The fossil imprint of a *Pikaia,* ancestor of the modern amphioxus (bottom), now considered the point of transition toward vertebrates.

cylindrical "backbone" onto which the segmented muscles attach (in herringbone fashion).

With time, the notochord will evolve into a spinal column, giving rise to that supporting structure which will one day allow land animals to walk, jump, and fly.

The creature now emerging from the mud is an ancestor of the amphioxus, generally considered the turning point in development toward vertebrates. This, then, is a very distant ancestor, but it will have to go through an evolutionary journey of over half a billion years to reach us.

With one short, rapid movement, the *Pikaia* disappears into the mud, leaving only its head exposed, its fine tentacles and appendages ready to capture prey.

Increasingly complex machines

It is time to reemerge. Drawing alongside the huge underwater cliff, we start to ascend, illuminating the vertical wall with our flashlights as we go.

Exploration of the Cambrian seas has been a real surprise. In a relatively short time span, evolution has diversified into an incredible range of animal life. Some of it will disappear, but some of it will give rise to a long line of descendants. And all of it has participated in the "revolution" which represents a watershed in the evolution of life: the birth of pluricellular structures.

This was already mentioned in the previous chapter, but it is worth repeating here because the fauna we have just encountered are very representative (much more so than hydras or jellyfish) of this important turning point. The crucial transition is from individual cells to superorganized cell colonies enclosed in an outer coating and endowed with a very efficient division of labor.

In the primordial seas, the passage from unicellular to pluricellular probably took place innumerable times, in various directions, and with more or less positive results. But what we see now are perfect machines, with a great variety of forms and innovative biological inventions, capable of dealing with the challenges of an increasingly competitive environment.

The change that is taking place in this period is comparable to the transition in the human species from a hunter-gatherer society to more organized societal forms, in which greater specialization and division of labor allowed for the increasingly efficient performance of the whole (guaranteeing each individual better survival).

Like primordial unicellular beings, hunter-gatherers started to associate in more and more complex ways. Here, the advent of pluricellular structures has changed everything: each cell has its own role, like an individual in society. For example, it can no longer procure its food alone because another individual is responsible for that (various cells gather it, digest it, and deliver it through the circulatory system and eventually eliminate waste). The transition is from a subsistence economy, in which each individual has to carry out all tasks, to a market economy in which each cell receives what it needs in exchange for something else.

This is the beginning of cooperation and specialization, partnerships galore between the blind man and the cripple (to return to the old analogy). This will lead to organs and parts which could not exist on their own but which can, as part of an interconnected system, provide precious services for the survival of the whole: movement, perception, digestion, excretion, duplication, and so on. That is to say: muscles, the nervous system, the stomach, intestines, kidneys, and genital organs. And, of course, the brain.

The first cerebral tangles

It is, in fact, in these seas that the first trace of cerebral organization appears: The nervous system, which was no more than a rough network in the jellyfish has, with the emergence of the ganglia, become an organized structure in the animals we have just seen. For the moment they are simply tangles of nerve cells situated in the heads of these primitive animals. But the sense organs—for sight, smell, taste, and hearing—are also starting to concentrate in the head (the most forward and exploratory part of the body).

Here, too (as in human society), only the advent of cooperation can allow for the emergence of new and very specialized functions. No nerve cell can survive on its own; it would die immediately. But set in a pluricellular structure, it can provide invaluable signals for the survival of the whole. Even though it is incapable of producing, digesting, and purifying food, it can produce behaviors—something that will become increasingly important for the welfare of the whole.

Therefore, cooperation has won out in the seas of the Cambrian period (as in those of the pre-Cambrian period): organized cooperation, in which every cell depends on the activity of the others and in which—a very important point—the malfunctioning of one cell threatens not only the survival of that individual but of all the others as well.

A new factor of selection affecting both individuals and the whole has come into play. It is not enough to perform well individually; all members of the team must pull their weight or the team will lose. In modern society this is known as "total quality."

Another factor has accelerated selection (and therefore diversification) in this new stage of life: the success of sexuality and death. Sexuality and death had already appeared, but in the Cambrian period their effects have increased exponentially.

The mixing of chromosomes resulting from sexual reproduction offers natural selection an ever vaster range of "card games" to play: the mechanism of dominant and recessive genes multiplies the combinations and enhances the variations. And it is upon this diversity that evolution is based.

Death and mass extinctions opened up new niches for more suitable living forms, providing them with "natural laboratories" for new biological experimentation. The Cambrian period sparked the diversification of life that was to provide evolution with new momentum. It also prepared the changes that were to lead to the conquest of dry land by very evolved forms of life.

Landing structures

While oxygen continues to accumulate in the atmosphere and the ozone shield (which will finally allow life to move onto dry land after three and a half billion years underwater) is forming in the upper reaches, the oceans are rapidly—albeit unwittingly—preparing certain "vehicles" suitable for the landing.

For life, the move onto land will be like changing planets: not only a different medium, but different gravity as well. The move can, in fact, be compared to moving from the moon to the earth, only it is much worse.

Take jellyfish, for example, so light and mobile in the water; on land, they are a shapeless mass. The same sort of thing would happen to astronauts if they were to land on Jupiter or Saturn, huge planets with much stronger gravitational fields than that of Earth: their bones would break like toothpicks and they would slump to the ground like jellyfish.

To leave the water and move successfully on land (let alone run, jump, and fly) requires a system of support capable of standing the enormous difference in gravity.

The Cambrian seas already offer a wide variety of "assembly lines" capable of producing vehicles suited to this change and many of the evolutionary lines that will be successful on land.

During our most recent immersion, we noticed two of these lines. One is that which will lead to crabs, scorpions, spiders, and insects: the "exoskeletons"; the other is the one that will lead through fish to amphibians and will finally give rise to reptiles and mammals: the vertebrates.

A twisted shell

As we ascend the face of the underwater cliff after our long exploration of the sea floor, the sun's rays start to filter through and warm the water.

The marine environment is completely different here and the diffused light gives us a good view of a submerged plateau. The colors are brighter, with the sea's blue contrasts with the red of the sponges and the violet and yellow of the actiniae swaying gently in the water. This small primeval garden is evidently very attractive to the numerous trilobites exploring it with their antennae.

Many of the other fauna inhabiting these shoals are also attempting to build "hard" structures (we already saw some armor while exploring the base of the cliff). Which of these animals will take the right road to land?

Around us are armored mollusks of all kinds. It's hard to believe that such a variety of shapes could have arisen from the basic idea of a worm building a house for itself. Yet, approximately 128,000 species of mollusks (including those that are extinct) are known today.

Many of the most primitive types inhabit these shoals. Will they ever be able to emerge from the water? For example, is that shell with its heavy protective armor wedged into a rocky crag really suited to life on dry land? It doesn't look like a good candidate.

There are, however, many ingenious shells around. One has a beautiful elongated spiral shape. Another magnificent construction looks like a tiny tower of Babel with its spires narrowing toward the summit: this is the *Murchisonia.*

Its spiral shape is not the result of an aesthetic quirk; it is a functional necessity. As so often happens in evolution, an initial defect has turned out to be an advantage: it is now believed that the shape is produced by a muscular asymmetry in the animal which causes a 180-degree rotation of the viscera at the larval stage, twisting the digestive tube into a U-shape and crossing some nerve bundles. But this warping accidentally has improved certain functions, resulting in appreciable advantages for larval development and probably new adaptations for respiration.

Which of the other mollusks that we can see here are suited for landing? A very strange-looking animal, a kind of jet, passes. Resembling a tiny octopus peering out of an ice cream cone, it moves swiftly through the water in the same way as a cuttlefish. The cone, which serves as a shell, is long and tapered: this torpedo is certainly well adapted for navigation in these seas but it doesn't look suited to continental life.

Toward a flexible exoskeleton

What is a *Chiton*? It is a very interesting mollusk for paleontologists.

Although it has a heavy and cumbersome plate of amour and moves with difficulty along the sea floor (it moves on its flat and gummy underside, like a self-propelled mattress), the *Chiton* is thought to lie halfway between various types of mollusks and could be a transition form to others. At the moment, very little is understood about its genealogy.

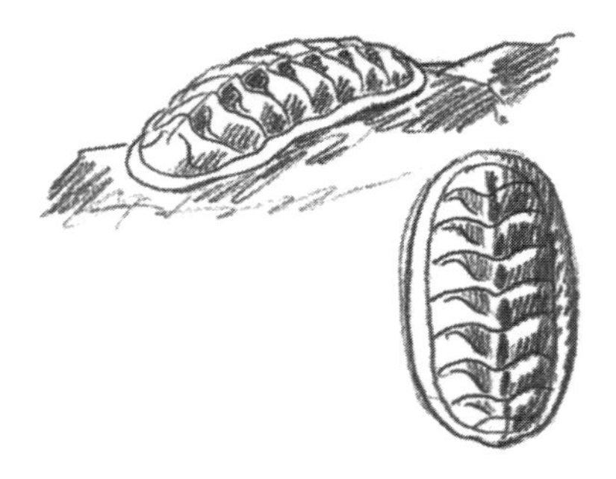

The same is true of *Pilina,* another mollusk with a rigid hat (sometimes in the shape of a "sugar loaf"). Like the *Chiton,* the *Pilina* is something of an evolutionary mystery. It seems, however, to have transverse links with and ramifications to other species of the Cambrian period.

The ancestors of the landing troops must be sought elsewhere: among the animals that have developed lighter and more flexible structures. "Worms," for example, have equipped themselves with coatings and supports that are more suitable to the changes occurring in the environment. One of their main features is their "metameric," that is, segmented structure.

Many of the animals found in the Cambrian seas have taken this path. And we have already seen some of them. The trick is to repeat one basic element throughout the entire body.

It is evident, for example, that a simple polychaete, worm, or centipede is made up of repeated segments. Individually, each of these has its own ministructure. In a polychaete, each segment has two "limbs," two nephrites (that is, tiny excretory organs), a pair of neural ganglia, some muscles, and so on; only certain organs, such as the digestive tract, the circulatory system, and the nervous system are common to all segments.

This structure is light and flexible, as well as resistant. It is undoubtedly far more suited to leaving the aqueous environment than a heavy plate of armor.

Toward the landing

Yet, an exoskeleton is not enough either. It takes various other preadaptations to make the conquest of dry land possible; for example, a system preventing or at least limiting the dehydration of tissues in the open air (such as a cuticle, that is a "wet suit" made of elastic but horny tissue), and a bent for wandering (it is unlikely that anything living in the mud could be suited to such a venture).

The group of creatures that seems to meet these requirements best is the arthropods. We met a few of them, in particular, trilobites, during our last exploration, but trilobites will never leave the seas. Centipedes, millipedes, and scorpions will do much better over the next millions of years, developing the anatomical and physiological qualities needed to land. We will take a closer look at them in the next chapter.

In the meantime, the other great evolutionary line—that of the vertebrates—is also developing. But it is developing more slowly. No vertebrates have appeared yet. It is the 15th of November (505 million years ago) and the only representative of the vertebrate line is the *Pikaia* that we saw at the foot of the cliff. It and other members of its group already have a rigid cord running down their backs, a precursor of the spinal cord.

Thus, on November 15th, things are gradually getting organized for the great adventure. On the next leg of our journey, we will finally shift to dry land to see the great protagonists of life on Earth emerge from the water, one by one. But neither the vertebrates nor the invertebrates will be the first on shore. The first to land are plants.

"Actually this is not a forest at all. It is a tuft of moss, . . . one of the first forms of plant life to develop on dry land."

6

The Landing of Plants

NOVEMBER 22ND (430 MILLION YEARS AGO)

Inside some moss

What a strange forest this is, so dark and damp! The green arbor above our heads is so thick that almost no light comes through.

Underfoot, the ground is soft and moist; there is not a blade of grass or a shrub anywhere. Surprisingly, all the trees look alike, smooth and apparently rubbery: no branches, just dagger-shaped leaves, like those of an agave, covering the trunk from bottom to top.

Actually, this is not a forest at all. It is a tuft of moss. We have reduced size once again to be able to get a detailed view of this plant and understand its structure.

Moss was one of the first forms of plant life to develop on dry land. It did not actually make the leap itself: it is only a descendant of some other plant form that developed the adaptations

needed to live permanently out of water after having spent a long time between water and land.

Some of these adaptations are evident on the moss's "trunk," for example, a kind of protective waxy cuticle to keep humidity in. One of the problems involved in the transfer to land is maintaining the aqueous environment needed for metabolism. The wind, the sun, and the air tend to dry and dehydrate everything that is exposed to them. Hence a skin is needed to preserve internal liquids (human beings have adopted the same solution: a skin that is both impermeable and porous).

We can see microscopic pores here, too, the so-called *stomia,* air vents that take in carbon dioxide for photosynthesis. As with cyanobacteria, the sun provides the energy for this process, allowing the plant to break down water molecules and recover the necessary hydrogen.

The mechanism of photosynthesis has remained exactly the same over the years: hydrogen and carbon are assembled to construct the tissue of the plant, while oxygen is eliminated as a waste product.

Water is abundant here because moss, with its cushion shape, retains all the water that comes its way: atmospheric humidity, rain, the spray from waterfalls and waves.

Let's climb one of these tiny "trees" and take a look around.

The climb is easy enough, as the blade-like leaves are rigid and form a kind of staircase. The need to capture as much sunlight as possible has led to a profusion of these ramifications (not branches). Large drops of water are nestled everywhere. This is particularly important since the plant does not have real roots.

When we get to the "tree" top, the view is completely different. Below us is a rocky precipice: we are on a large rock overlooking the sea and a saltwater lagoon. To one side, waves splash onto the shore, sending fine spray all the way up to us. All around, rocks form a sheltered area. Just a few yards out, a tangle of algae can be seen floating gently back and forth like hair blowing in the breeze.

Algae, ancestors of plants

These algae are large pluricellular structures, distant descendants of certain primitive eukaryotes that engulfed and captured cyanobacteria in order to be able to use sunlight to build sugar molecules that could, in turn, be used as fuel. In a certain sense, their evolution has been similar to that of animals: marine algae include the equivalents of sponges, worms, and other organized forms. These algae are huge colonies of photosynthetic cells that are well adapted to their habitat.

There are numerous varieties of algae in the modern world: some small, some large, either entirely submerged or buoyant, with colors ranging from intense green to yellow, red, and brown. Some can live at depths of up to six hundred feet, while others fasten onto rocks at the water's edge. They are so different that the term "algae" does not have systematic value in botany; it is used to indicate the many plant forms that lack specialized parts and consist only of a thallus.

The thallus, which is the basic plant structure, can take different shapes: a leaf, a ribbon, a float, and so on. In these algae, the pseudoleaves are called "laminae" and are basically tiny solar panels. Some of them reach a dozen yards in length: one kind of alga, the *Macrocysti,* can grow up to a length of 450 feet, the longest known form of life.

The thallus has one surprising characteristic: any piece of it can reproduce the entire plant (as if a severed human arm could reproduce an entire human being). Regeneration by one part is not typical of plants alone; there are similar cases in the animal world. For example, if planarians, fresh water worms, are cut into pieces, each fragment will regenerate a new individual. A piece of the tail can regenerate the head and vice versa.

Of course, the shape of algae has also been modeled by the effects of gravity, which is much weaker in water. Thus, it has produced a very light structure that is totally unsuitable for life on earth. In fact, the few yards lying between the algae and dry land constitute an unbridgeable gap separating two worlds: different gravities, different environments, different rules of nutrition and reproduction.

Yet, certain small algae were the first to accomplish the definitive changeover from water to land. The moss that we are sitting on now is one of their descendants.

How was this monumental change achieved? This may be a good opportunity to review what we have found out about this exceptional adventure.

It is possible (even though there is no proof of it) that the first forms of life on Earth were simple one-celled organisms capable of photosynthesis. That is, they may have been very adaptable and resistant unicellular algae (actually bacterial cells) carried onto dry land by the wind, rather like the landing of paratroopers.

Cyanobacteria of this kind still exist on Earth today and manage to survive under extreme conditions (some proliferate on the snow at over 15,000 feet altitude and have special pigments against ultraviolet light which give the snow a reddish tinge).

Lichens may also have been among the first multicellular eukaryotes to adapt to life on Earth, thanks to the simplicity and adaptability of their structure. As already mentioned, lichens are an association of two types of lower eukaryotes: green or blue algae and fungi. The algae provide the system with energy; the fungi provide the humidity and protection against dehydration, and the attachment to rock.

There are no fossil remains of lichens or any unicellular algae that prove that they were the first forms of terrestrial life (if only by a short time). But even if they were, their transition would probably not have had the consequences that the landing of more evolved forms had: marine algae, for instance, were already well-developed plants with a pluricellular structure—the "green" rulers of the seas.

Five hundred million years ago, these plants were already very close to the point of landing. What difficulties did they have to overcome to pass from the aqueous to the terrestrial environment?

The four problems involved in landing

There were four basic problems to be overcome: (1) protection against dehydration, (2) support, (3) nutrition, and (4) reproduction. These are the same problems that animals later had to face during their conquest of dry land. The fifth problem, ultraviolet radiation, had been solved by the ozone layer which was already effective 500 million years ago (some people feel that the present oxygen concentration in the atmosphere was reached 400 million years ago).

Let's see how the algae solved these problems, because the story of these solutions is the story of the arrival of life on land.

Before attempting to change worlds, the most favorable place for marine algae was just below the tideline, in shoals shallow enough to allow them to capture solar energy but deep enough to keep them covered during low tide.

What probably happened is that algae began to find themselves in the risk area from time to time, that is, in the area left exposed at low tide. As a result, they started to develop adaptations to tolerate the air. Thus, they became something like the opposite of scuba divers: capable of surviving for a short time in the terrestrial environment before returning to their natural habitat—the kind of alternation typical of amphibians.

Straddling two environments allowed the algae to take advantage of both: for example, they could use the solar energy and carbon dioxide provided by the atmosphere, while relying on the water provided by the tides to avoid dehydration. Another advantage of remaining outside of the aquatic world for even only a few hours was being out of reach of herbivorous predators.

At a certain point, selection of the most suitable forms allowed for development of "amphibious" algae, which had a cuticle capable of keeping moisture inside. It also led to the development of stomia, that is, pores, in the cuticle, allowing for the exchange of gases with the intake of carbon dioxide and the release of oxygen (exactly the opposite of what happens in our lungs).

The algae, therefore, started to live with their feet in the water and their heads in the sun, so to speak. But landing could only take place once the problem of adequate water supply and support systems had been solved.

Obviously, only very small plants could overcome this problem easily in the beginning. The moss we are sitting on was one of these initial "landing troops," meaning that it was one of the first plants to leave the aqueous environment forever. It may have passed to a lagoon first, though, to adapt to brackish water before passing on to fresh water, thereby learning to defend its internal saline environment.

In fact, this moss still has the ocean inside it in the form of the liquid trapped inside its membrane and protected by the cuticle. Its water supply is provided by environmental humidity and rain.

It does not have an exoskeleton yet or a real supporting structure. A broken "trunk" not far away reveals the lack of any support allowing for real vertical growth. But moss "stands" nevertheless. The trick is to remain low and close to other individuals so that they support one another: rather like a wall-to-wall carpet, in which each fiber would not be able to stand up if it were alone or a yard high.

Only later will other forms of plant life arise that are endowed with far more rigid structures capable of vertical growth thanks in part to the systems for internal transportation of liquids.

These other forms will be discussed shortly. In the meantime, let us get back to our moss.

Cannonballs of water

A strong wind has come up and clouds have moved in, darkening the sky. The temperature has dropped. A rainstorm is on its way.

A bolt of lightning crashes to the ground not far away, the thunder explodes in our ears. Torrential rain starts to fall. Drops as big as cannonballs pound the moss we are sitting on. We try to seek shelter under the leaves, but gushes of spray splash up, soaking us to the bone. The moss "trunks" vibrate under the violence of the downpour. At our size, this rainstorm is like a hurricane.

The water dripping off the leaves and the "steps" accumulates on the tangle of filaments at the base of the trunk. Protected from the sun, it will remain there, forming a reservoir for later use.

Naturally, water is vital for photosynthesis: no plant can survive without it. But in this stage of life, water is also indispensable for another basic function: reproduction. That was the last problem that had to be solved before plants could leave the aquatic environment. And moss has accomplished it.

In the water, it is very easy for plants to disperse their male sex cells in search of female cells to fertilize. The currents are an excellent means of transportation and reach all corners of the sea. This is not the case on dry land, however. Sperms can swim, but they cannot walk. Plants, therefore, had to find a different way to disperse their sex cells.

This is where the providential downpour comes in. It serves as a nuptial conveyance: carried by the water, the sperm circulates in search of the eggs of other plants. And that is surely happening right now in the water streaming down from the moss. Attracted by the chemical substances emitted by the egg cells, the male cells are traveling to fulfill their mission—fertilization—thereby achieving the fundamental aim of reproduction.

Sex and water

But reproduction is not as direct in these primitive plants (and animals) as it is in human beings. There is an intermediate stage. That is, the sperm and the egg give rise to an intermediate "individual" (in this case a long stem surmounted by a capsule full of spores that will pop open like a bottle of champagne at the right time) which disperses the spores from which the new plants of moss will finally germinate.

This system has a number of advantages but also one major disadvantage: fertilization still requires water. This is a serious drawback to the diffusion of plant life on land because it means that plants cannot move away from the wetlands. Eventually, the development of seeds (parallel with that of plants more suited to drier climates) will eliminate this intermediate step. And the creation of separate male and female seeds will overcome the hermaphroditism of the first stages of plant sexuality.

The terrestrial evolution of plants (which is in many ways astoundingly similar to that of animals) will take place in a relatively short time compared to the time it took life to evolve in water. In only 100 million years (between 450 and 350 million years ago), that is, in just over 1 percent of the existence of life on Earth, plant life will pass from the first mosses to the diversity of the Carboniferous period, with its luxuriant forests and towering trees.

Three basic developments will allow for the autonomy and vertical growth of plants: roots (making it possible for plants to draw water and mineral salts from the ground), cellular thickening (permitting the growth of woody layers and therefore trunks), and internal canals (for the rise of liquids and nutrient substances by osmosis).

Thanks to these simple but efficient devices, plants will soon cover all land surfaces, gradually gaining height to compete with nearby plants for a "place in the sun." At a certain point, leaves will develop so that more solar energy can be absorbed by the tiny solar panels.

Unintelligent growth

Yet, the great success of plants and their extraordinary ability to produce food directly without having to prey on others will also delimit their evolutionary destiny: they lack movement and, in the end, intelligence.

In fact, no plant possesses anything even similar to a nervous system that allows for motion and perception of sounds, images, scents, and tactile sensations. It is true that plants can move (this can be seen in certain speeded-up film sequences) and that they are sensitive to light, certain chemical substances, and even touch (as some retractile leaves reveal), but these primitive perceptions are of a chemical—not a nervous—nature. What is missing is a center that memorizes, processes, and creates behavior.

The primitive network that we saw in jellyfish was the first sign of a fundamental evolutionary innovation marking a turning point toward a completely different line—one that would lead to increasingly mobile, perceptive, and intelligent forms of life.

In the 300 million years of their existence, plants have not shed their evolutionary yoke: their initial moves, which were certainly successful, were too specialized to allow them to change direction and evolve in different ways. But it is thanks to the road taken by plants that the earth became a hospitable place. Their diffusion on dry land created an entirely new environment brimming with food and resources where only arid lava deserts once existed—an environment suited from the beginning to animals and, therefore, to their emergence from water.

"It's an enormous millipede. Its legs are perfectly synchronized in a wave-like movement that rolls from one end of its body to the other."

7

The Landing of the Invertebrates

NOVEMBER 27TH (380 MILLION YEARS AGO)

A one-piece electric train

An unfamiliar rustling is getting louder and louder. The noise is soft and padded, as though someone were walking on the minute vegetation around us.

No larger than insects, we are leaning against a small twelve-inch rush standing in the midst of a forest of slender branching stems. From our perspective, these "trees" cutting out the sun that slants in obliquely to illuminate patches on the ground look like giants.

As the rustling gets louder, we begin to hear a kind of crackling that is too regular to be accidental.

All of a sudden an enormous (from our point of view) millipede appears from behind a little knoll. Its legs are perfectly synchronized in a wave-like movement that rolls from one end of its body to the other. It is a most impressive sight: the animal resembles a one-piece electric train with ring-like segments that give it the flexibility of a coil.

The tentacle-like antennae protruding from the forepart of its head explore the terrain. The millipede uses its antennae as tactile and taste sensors to orient itself. They are basically two long, thin tongues for gathering information. Behind the antennae are two primitive eyes, each one composed of a group of tiny domes sensitive to light.

The animal stops and raises its head to look around, revealing the strong plates that form its mouth. They look like tiny claws and serve to cut food. Despite the impressive appearance of these jaws, we know that the millipede is a vegetarian and that we have nothing to fear.

Now we can see its segments more closely: each has a pair of legs on either side. Some millipedes have as many as one hundred segments, that is, a total of four hundred legs. Unfortunately, we don't have the time to count them, but the animal looks very long indeed.

This millipede is a totally terrestrial animal; it is the descendant of other millipedes which, somewhere around November 23rd (little more than three "days" ago), were the first animals to carry out the feat of passing from water to land.

How to breathe?

We have no way of knowing exactly how things went, but we can imagine that animals not very different from this one (and that once lived close to shore) started to explore dry land.

Land provided numerous advantages: food, new niches, no predators. But it also meant adaptation to a completely new environment and required new solutions to two very important problems: respiration and reproduction.

Breathing on land called for a new way of absorbing oxygen. Practically, it meant changing from a system like that used by trilobites ("feathered gills") to one more suited to the terrestrial climate (which would rapidly have dried gills out).

The new mechanism, perfected in the course of a long amphibian-type evolution halfway between water and land, is in plain sight on this millipede's "sides." (It is very similar to the solution adopted by plants: stomia). Each segment has pores, which act as air vents conveying oxygen to the cells through small branching channels. This makes each segment independent, with its own legs, muscles, and respiratory apparatus.

A very simple but effective solution. And one that permitted this animal to leave the aqueous environment without having to develop a large, complicated respiratory system. Indeed, the solution is so appropriate that it still functions perfectly in modern millipedes.

Millipedes may have landed about 400 million years ago: the oldest fossil findings date back to that time. But millipede fossils from about 390 million years ago are abundant, indicating that they were widespread and enjoying rapid and increasing success at the time.

Their genealogy, however, is still unclear. They are probably related to trilobites (which look like flat millipedes) even though they are not direct descendants. Some think that both organisms originated from a common ancestor: a segmented marine worm, probably the forefather of several "varieties" of invertebrates.

Primordial intercourse

An unexpected sound makes us jump: the millipede has darted off toward a dark object half-hidden in the vegetation. It takes us a while to realize that the motionless object rolled up among the rushes is another millipede.

The two animals meet and begin a strange ritual. The surprise visitor is probably a male. No doubt he was attracted by certain chemical substances released by the female and picked up by his antennae.

This is a mating ritual: the first, perhaps, in the history of life on dry land.

As already mentioned, one of the most crucial problems posed by the transfer to dry land is reproduction. And the reason is the same as for plants: the absence of water. In the seas, the currents and the water favor the transportation and encounter of sperm and egg cells. On dry land, this does not occur; the force of gravity pulls everything to the ground (we saw that the mosses use dripping water to transport their gametes).

Millipedes have developed another, far more effective solution. They deposit their sperm directly into the body of the female. In this way, they enjoy various advantages: first, they solve the problem of casual encounter, which is very difficult on dry land. This method provides delivery to the door. Second, they solve the problem of the right time for fertilization: the male's spermatozoa are not used immediately. They are enclosed in a sac and preserved under ideal conditions, as if in a tiny "aquarium," where they remain "on call" for fertilization of the eggs when they are deposited on the ground.

It is important to point out that all the animals that left the seas maintained an aquatic environment in their bodies and in each of their cells. And that the concentration of salts in their body fluids is surprisingly similar to that found in the seas. This is the direct legacy of their original habitat.

The two millipedes are now clasped in a sinuous embrace: their dark, shiny bodies move slowly, legs sticking out everywhere.

It's hard to make out exactly what is going on, but the process is probably very similar to the mating of the millipede today: the male deposits his sperm into the body of the female. There is no penetration, however: fertilization takes place "by hand." The genital organs of both the male and the female are located at the base of their second pair of legs. The male uses his seventh pair of legs to place the sperm sac into the female's cavity.

As mentioned earlier, the female will only deposit her fertilized eggs when she finds a suitably protected place.

The first predators

While the millipedes are still united in their strong embrace with the male on top busy transferring his sperm, a completely new form of life passes: a spider.

Spiders belong to a subsequent wave of "landing troops." But their segmented structure suggests that they share a common ancestor with millipedes. Fossils reveal, in fact, that body segmentation is much more pronounced in this primordial spider than in modern spiders.

With eight legs resting on the ground like a lunar module, the spider is certainly agile and stable on land. Its admirable system of independent legs allows it to gain a good footing on even the most irregular terrain. It is amusing to think that NASA's engineers have often been inspired by this creature when designing self-propelled vehicles to be sent to other planets.

Its belly is probably already developing the gland that will be able to secrete the liquid used for spinning webs—a liquid that turns into a fine thread upon contact with air. But the time for spider webs has not yet come: no insects are capable of flight and there is no reason, therefore, to prepare such a trap.

Spiders may have used their threads in primordial times to tie up and immobilize prey. The black widow still adopts this tactic today. But our millipedes are too big to be attacked and the spider continues on its way, marching through the vegetation in search of other victims.

"With eight legs resting on the ground like a lunar module, the spider is certainly agile and stable on land."

A nocturnal attack

Night has fallen. The miniature forest in which we are observing the first scenes of terrestrial life is illuminated by a full moon, pale but nevertheless bright enough to create areas of light and shade and to provide light for us to see.

We have taken shelter on a small horsetail to escape the fearful predators that have started to roam the land. The presence of undefended fauna has created an excellent "market" for strong and aggressive carnivores. Therefore, sleeping on the ground is not a good idea, whether by day (daytime predators include spiders) or by night. But we did not intend to sleep tonight in any case: caressed by a gentle breeze, we are waiting for the arrival of that ferocious nocturnal predator, the scorpion.

The land scorpion descends, even if not directly, from similar forms that existed in the seas (some of which reached six feet in length). Its adaptation has gone through the same steps taken by other landed animals to adjust their exoskeletons, muscles, and reproductive techniques to the requirements of dry land.

In the scorpion, the respiratory problem has been solved by converting the gills into a closed structure, that is, a few lamella-filled pockets in the abdomen. The air pumped in from the outside by a muscle acting like a bellows circulates in the lamellae, where it is brought into contact with the blood vessels. It is a prototype of our lungs.

No one knows for sure when scorpions stepped onto land. The fossil record shows their presence much earlier than spiders, probably as long as 410–400 million years ago. But it is unclear whether those scorpions were adapted to living on land permanently or only temporarily.

They must have landed not long after the millipedes. In fact, millipedes had to wait for the presence of plants to provide food and scorpions had to wait for the presence of millipedes (and other invertebrates) to provide prey.

And there it is, finally! A dark shadow with two pincers and a raised tail stalks through the vegetation. This beast is big and black and has that awful appearance that we are all familiar with. It almost slides forward, as if it were pushing a cart of some kind. Its erect tail clearly reveals the segmented structure of its body.

It stops. It may have perceived the presence of prey. If the scorpion detects the male millipede that is still in the vicinity, the latter's life will not be worth much.

The scorpion moves suddenly, then stops dead again, rather like a hound pointing fowl. Its tail is poised; it is clear that it is preparing an attack.

Suddenly it dashes off, disappearing behind a small rise. All

"And there it is finally! A large shadow with two pincers and a raised tail advances through the vegetation. This scorpion is large. . . ."

we can see is the tip of its tail. A rustling breaks the silence of this memorable Paleozoic night. The scorpion darts forward again, now vanishing completely from sight. After a few dull thuds, the millipede's tail writhes through the air.

Rolling into a ball is not sufficient defense against a scorpion. The millipede's tail whips back and forth once again, then the thuds diminish and finally die away. A tiny primordial drama has just ended, as it has millions of other times, without a cry, without a trace.

The millipede has made its contribution to evolution, however: in the form of food in this case. But it has also done its part as a reproducer; its sperm is intact in the female's body, like an inheritance in a safe. In the relay race of life, descendants will in turn transmit their genes (mixed and transformed) to successive generations for millions and millions of years to come.

We are approaching the end of November and the earth is starting to take on a different appearance: plant life, although still miniature, is beginning to paint the coastlines, the lagoons, and the riverbanks green.

Like a seed sprouting and bursting through the ground, the seed of life, after years of incubation underwater, is germinating from the sea and spreading to dry land. The first two landings—that of plants and that of invertebrates—have already taken place. Now the most important one for the subsequent evolution of life is about to take place: the landing of vertebrates.

So let's go back to the sea for the last time to follow this great event.

"Here we are underwater again. . . . Huge nautilus-like creatures jet by in the distance with their cone-shaped bodies. . . . But there are also new forms to be seen. . . ."

8

The Protagonists Arrive

NOVEMBER 27TH (380 MILLION YEARS AGO)

A living museum

Here we are underwater again, exploring a world that has changed immensely since we left it at the end of the Cambrian period over 120 million years ago.

Surveying the sea bottom on this 27th day of November is like visiting a natural history museum with a vast and rich collection. On display before us are the actors of the first part of the story of life, the various "passengers" that got off the train during its long evolutionary course.

Slipping among rocks and over sand flats and algae meadows, we see many of the creatures that we have already encountered on our previous dips (or rather, their descendants, which are substantially similar).

The waters are, of course, still inhabited by prokaryotes,

cyanobacteria, and stromatolite formations, as well as by protozoa of all shapes and sizes, sponges, jellyfish, worms and marine snails, gastropods with their shells, and trilobites. Huge nautilus-like creatures jet by in the distance with their strange cone-shaped bodies and long mobile tentacles. They have now reached immense proportions and some are yards long.

But we also notice new forms of life on the sea bottom such as starfish and ancestors of the sea urchin (which only appeared later). We decide to take a better look because, as strange as it may seem, these are our distant relatives, at least insofar as they are an early offshoot of the evolutionary line destined to give rise to the chordates (animals with a spinal cord which will eventually evolve into a spinal column). More simply, they share a direct common ancestor with the evolutionary line that will eventually lead to fish, amphibians, reptiles, and mammals.

Like sea urchins, starfish seem to have little in common with this line. They don't even look like animals and yet their internal structure exhibits the typical bilateral symmetry of vertebrates (humans also have it: the left-hand side of our body is the mirror image of the right-hand side). Their radial shape was a subsequent exterior adaptation. (It is interesting to note that a starfish looks like an urchin if it is closed into a ball like a hand closing and, likewise, that an urchin resembles a starfish if it is opened up.)

The starfish is now slowly moving, thanks to an internal hydraulic system which allows it to advance on its "ambulacral peduncles," that is, a series of tiny mobile protuberances on its underside. It has a fragile inner "shell" which may be considered a kind of skeleton as it is totally enclosed by the colorful skin.

But it is the fate of the starfish (and the sea urchin) to remain at this stage forever, with very little or no further evolution. Like so many other forms of life, it has ended up in a cul de sac—an evolutionary path that does not lead on to others.

Toward fresh waters

Our musings on the fate of these animals continue. It is true that they have gone into evolutionary "retirement," meaning that they will go on duplicating themselves without much variation for millions of years to come. But it is equally true that this is also a manifestation of their great adaptive success, since only a formula that is successful in its environmental niche can survive for so long. While our thoughts drift on, a strange elongated, snake-like shape appears in the distance. This animal is unlike anything we have seen so far.

Curious to find out more about it, we decide to follow as it heads for a rather turbulent stretch of the coastline, where churning sand and other particles keep visibility low. The temperature of the water is lower and we can feel a slight current flowing against us: this must be the mouth of a river.

The water is becoming increasingly colder and more turbulent and our mysterious animal has disappeared. But it has guided us to an environment—that of fresh and brackish water—which has become very important for the further development of life, especially on the eve of the emergence onto dry land of the future lords of the planet: the vertebrates. It is, in fact, from places like this that the most successful attempts are being launched to leave the seas and conquer the new territories that are now accumulating food and prey.

Armored vacuum cleaners

We have been swimming upstream for some time now. The water is no longer turbulent, but we still cannot make out the banks. We have probably entered a bay or lagoon. The water is

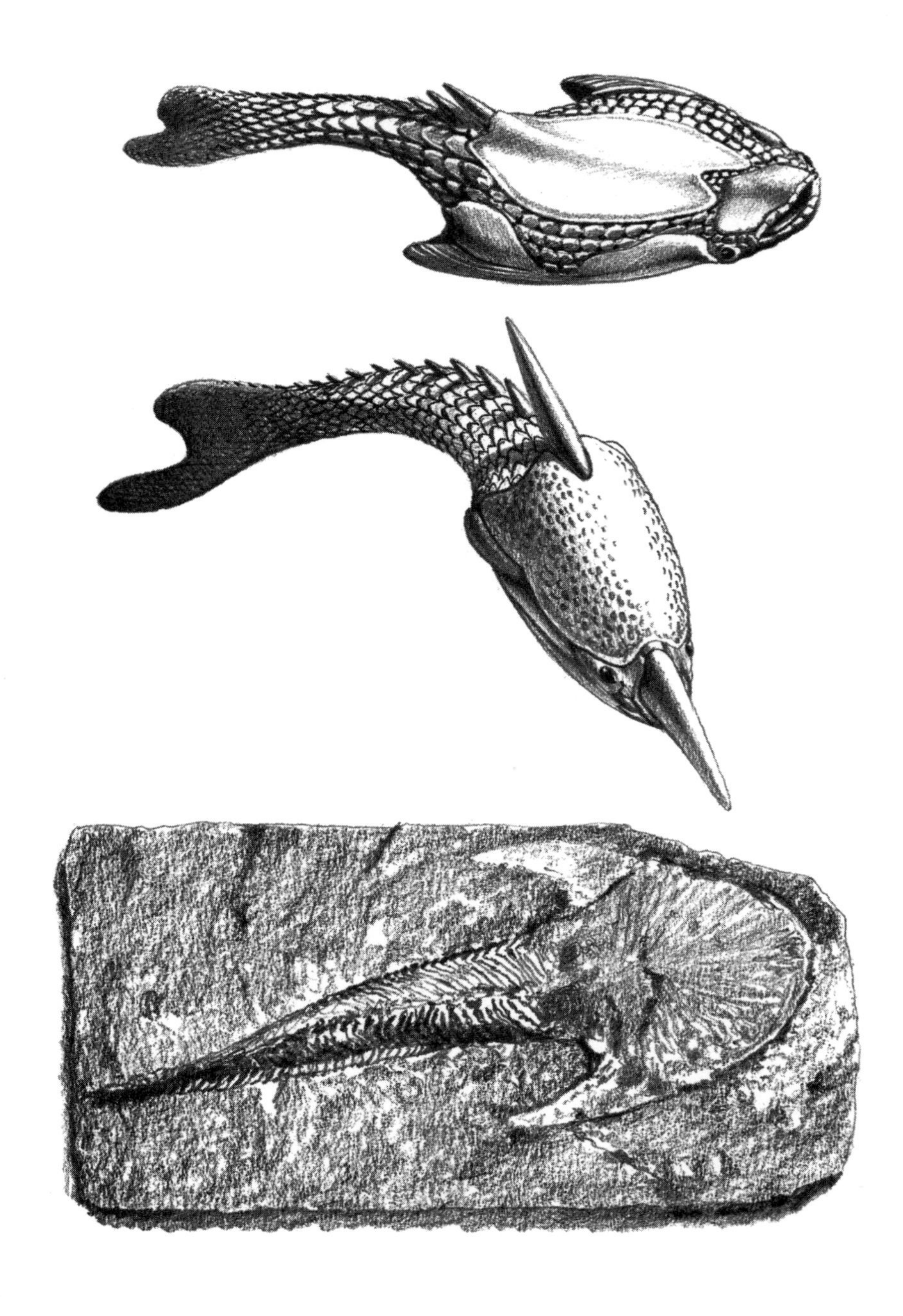

clearer and we can see the muddy bottom covered with tufts of gently swaying algae.

A strange triangular object is poking up from the bottom. From where we are, it looks like a spade with two studs or bolts stuck on it. But as we move closer, the "spade" suddenly shoots out of the mud and darts away. This is an ostracoderm, that is, a protofish, a very primitive fish with an armored head.

The only visible part while it was settled obliquely into the mud like a missile on its launching ramp was the head, an unusual rounded triangular shape. But we did manage to catch a glimpse of the body as it slipped away: it was slender and aerodynamic, rather like a gigantic tadpole.

This *Drepanaspis,* an ostracoderm about one foot long, is armored like a medieval knight. A broad bony plate—a kind of flat helmet—protects the top part of its head, while other smaller plates protect the front and sides. The eyes (the two studs) are lateral. The body is covered with scales and small spines.

Another protofish is slithering around in the mud, but this one is completely different. It has a single, steplike helmet on its head which terminates in small wings probably meant to function as stabilizers for swimming. The helmet is topped off by a long bony spine. But the snout is the most astonishing feature of all: its long point makes this ostracoderm look like Pinocchio.

This creature is called *Pteraspis* and is a proficient swimmer. Relatively small (no more than eight inches long), it is only one of the many ostracoderms that inhabit the fresh waters of the Devonian period (that is, the current period, named after the fossil deposits found in Devonshire, a county in Great Britain). These primitive fish have a number of characteristics in common: in particular, they have armored plates and do not have a jaw. In

---

Opposite Page: Bottom: this fossil imprint of a *Cephalaspis* which dates back 420 million years was found in the old red sandstone deposits of Scotland. Center and top: a *Pteraspis* and a *Drepanaspis,* respectively. Three small ostracoderms ranging from six to twelve inches in length.

fact, the class they belong to is known as the *Agnatha*, which means "without jaws." Thus, despite their awesome appearance, they are not predators. They filter water for food.

This is basically what whales do today on a much larger scale. The difference is that ostracoderms eat not only plankton (although some are specialized in eating it), but also and above all waste that is floating on the surface or that has sunk to the bottom. They are real "vacuum cleaners," capable of filtering great quantities of water and then expelling it through openings at the sides of their throats.

These animals also share another important characteristic: they are the first vertebrates. Although their skeletons are still cartilaginous, they already possess that extraordinary invention which is to be so successful in the course of future evolution: the spinal cord. How did this supporting structure arise?

The origin of the backbone

Not much is known about the ancestors of ostracoderms; in fact, there are no fossil remains to help us in this case. A few reasonable hypotheses can, however, be advanced. Some start with the *Pikaia*, the animal similar to the amphioxus that we met in the Cambrian seas, the first to exhibit a rudimentary spinal cord in the form of a notochord, that is, a thickening running down the length of its back.

Interestingly enough, a similar thickening appears during the embryonic development of present-day vertebrates (including human beings). The vertebrae arise along the thickening and later "choke" it off. This embryonic stage suggests a link between these two moments of evolution.

Fossils provide us with no clues about the links between animals such as the *Pikaia* and the first creatures with a backbone, but an intermediate form still exists today: the lamprey.

The modern lamprey is, of course, different from its ancestor which swam the primordial seas. Yet, many scientists feel that some of its very primitive features provide a good model of how evolution may have proceeded. In fact, the spinal structure of the lamprey seems to be halfway between the notochord of the *Pikaia* and the spinal cord of ostracoderms. The lamprey's notochord already has rigid structures (arched rods) which prefigure the spinal cord; there are no real vertebrae, but there is a rudimentary supporting column.

Ostracoderms do not necessarily descend from the ancestors of lampreys, but it seems reasonable to assume that the evolutionary "shrub" branched out into various lines of these animals at that time and that some of the branches gave rise to the ostracoderms.

The brain and the third eye

Can modern lampreys tell us something about the level of development of the nervous system of animals during the Devonian period? The marine organisms that we met in the Cambrian possessed a very rudimentary brain in the form of small nervous tufts or tangles. The lamprey, on the other hand, already has two tiny brain hemispheres with primitive lobes (that is, specialized areas) for sensing odors.

It may seem strange, but fish have a very highly developed sense of smell. Water is full of odors, or rather, chemical compounds that signal a variety of things, from the presence of predators or prey to harmful substances, food, and so on.

In addition to the olfactory sacs located in its nasal cavity, the lamprey has taste buds in its pharynx. It also has a primitive balance control system in its inner ear (like human beings) with semicircular canals that use a clever bubble mechanism to give the position of the head in space.

Did the ostracoderms and their ancestors already possess these features? The answer is probably yes (and it is significant that the fossils of fish that appeared soon afterward, the placoderms, have very evident nostrils situated in close proximity to their eyes).

The fossils of some ancient ostracoderms (such as the *Cephalaspis*, which lived over 400 million years ago) clearly display a third opening between the two eyes: the so-called third eye. This area sensitive to light was linked to the pineal area of the brain and probably operated like a photoelectric cell, signaling day and night, swimming trim, and perhaps even the shadows of prey arriving from above. The lamprey still has this third pineal eye endowed with a clear lens and a pigmented retina.

Dissection of the cranium of certain ostracoderm fossils has revealed the structure of the cavity containing the nerve bundles and blood vessels. The brain of these animals was much like that of modern lampreys, including the system of equilibrium composed of two semicircular canals.

It is interesting to note that the lamprey is still an animal which alternates between two environments: like the eel, it commutes between the sea and fresh water. It may have been an animal like this that brought some ancient species closer to the big leap from water to land.

The giant scorpion

Two ostracoderms suddenly flatten out and dig themselves into the mud, disappearing in a small murky cloud. Something has obviously frightened them: have they perhaps perceived the approach of a predator? These animals, like lampreys, probably have some kind of system that is sensitive to changes in water pressure. This is a kind of perception—somewhere between

"The water scorpion is almost six feet long. . . . The ostracoderm writhes in an attempt to free itself, . . . but its armor is not sufficient to defend it against a devil this size."

touch and hearing—with which we are unfamiliar, but almost all kinds of fish have it. Generally situated along their sides, it is called the "lateral line."

And here comes the predator: a huge water scorpion, moving silently through the water like a tank with its powerful pincers outstretched in search of prey. Almost six feet long, the scorpion uses two small limbs and its tail to propel itself through the water (like a dolphin).

Now it settles onto the sea bottom. It must have sensed the presence of a prey because it advances slowly, moving its eight legs in sequence. The powerful pincers start to burrow in the mud. There is agitated movement below: something is trying to escape. The prey darts forward. For a moment, it seems to have made it. But with a flick of its tail and its limbs, the scorpion pounces on the hapless creature, seizing it in its pincers.

The animal writhes desperately in an attempt to free itself, but the scorpion's grasp just behind the head is solid; there's no hope.

Now we can see the prey more clearly: it is a medium-sized ostracoderm. Its armor is just not sufficient in this situation: it may help against smaller predators, but not against devils this size. The scorpion tears it apart with its pincers and pops the pieces into its mouth.

This scorpion does not have a poisonous stinger on its tail. It is, in fact, not a real scorpion but a *gigantostracchus* of the merostome class of crustaceans.

The huge animal stalks away, clutching in its pincers what is left of the lifeless ostracoderm. The pressure waves created by this struggle must have been felt at some distance because another scorpion has hurried to the scene looking for its share of the feast. Sharks behave in the same way and are attracted to a banquet by blood.

The time has come for us to retreat and get on with our exploration of the rest of this underwater world now preparing for new and important transformations. It is in fact here in the seas of the Devonian period that a new structure is coming into being: jaws, the traplike mechanism that will give shape to some of the greatest predators of all time, from sharks to tyrannosaurs, crocodiles, and lions. But first it will give rise to the placoderms, the terrors of these waters.

NOVEMBER 27TH (370 MILLION YEARS AGO)

An armored monster

Before us is a genuine monster. Between twenty and twenty-five feet long, it circles around for prey like a shark, its immense jaws hanging open. It has no real teeth, only huge fangs more than half a yard long.

It vaguely resembles a killer whale in profile and size, but is much more terrifying: its head, jaws, and part of its body are covered with large plates of armor. This ferocious and implacable death-dealing machine is a predator of the open seas which, like the killer whale, occasionally enters shallower waters.

All forms of life in the vicinity have disappeared since *Dinichthys* (the name of this creature, meaning "terrible fish") made its appearance. All fish, large and small, have fled or are keeping their distance. Even water scorpions avoid this absolute tyrant; no one can compete with its strength.

The revolution that has taken place in the marine hierarchy is the result of a simple but decisive anatomic invention: the jaw. Until now, vertebrates did not have jaws, only a mouth, an opening with which they took in food. At best, the mouth was hardened around the edges, perhaps with some horny protuberances that acted as teeth. Never before had there been anything like an articulated jaw, activated by powerful muscles transforming the mouth into a steel trap.

The appearance of jaws

When did jaws develop? Fossils indicate that the first animals to have jaws as described here were acanthodes, small freshwater fish covered with bony scales that lived about 440 million years ago.

Before us is a genuine monster: a *Dinichthys,* a huge armored fish, twenty to twenty-five feet long, maybe longer.

Acanthodes were similar to many other kinds of fish evolving at that time, especially sharks (in fact they are called armored sharks), and to some bony fish and placoderms (the class to which the *Dinichthys* belongs). They formed a kind of "mixed group" that had vague links to various other groups. Given the success of the first experiment, jaws gradually extended through the labyrinths of evolution to all new emerging forms.

The placoderms, including the *Dinichthys,* developed this characteristic in a very spectacular way and rapidly obtained enormous advantages as predators to the detriment of the toothless ostracoderms. Placoderms of various shapes and sizes swim the waters of the Devonian period. But the *Dinichthys* is at the top of the predatory hierarchy.

But how did jaws develop? Fossils do not provide much material for a reconstruction. Yet, an understanding of the way things went may be gained from study of the anatomy of an ancient fish which is still living today: the shark.

Sharks, which first appeared 390 million years ago (160 million years before dinosaurs!), still circulate in our seas literally unchanged. Examination of a shark's jaws reveals that they have been formed by a transformation of the gills: the first branchial gill, or arch, has shifted forward and been incorporated into the mouth. The second branchial arch reinforces this structure, allowing for efficient articulation with the cranium.

Sharks' jaws are still not welded to the skull, as becomes obvious when they attack a prey: their jaws come out of the mouth, allowing their terrible teeth to sink in deeper. Only later and in other fish did the maxilla become fixed to the head and the mandible solidly attached to it by a bony joint.

It is curious to note that in mammals, residual gills (which are absolutely useless outside of water) transformed into other parts: the hyoid bone, which supports the tongue; the cartilage of the larynx (the Adam's apple), which hosts the vocal cords; and perhaps even the first tract of the trachea, whose various "rings" are probably left over from primordial gills.

Since the jaws had to be set with teeth, the Devonian period saw the rise of this other fundamental innovation. Teeth are believed to have evolved from the skin. The placoderms had developed very sharp-edged plates and even fangs by transforming and modeling the bony armor that protected their heads. But real teeth (like those of sharks) are thought to have arisen from an alteration of the bony scales that covered the body.

Sharks' teeth—arranged as they are in rows with those farther back ready to move up to take the place of those that fall out—strongly resemble the sharp, hard scales of sharkskin. Sharkskin is, in fact, covered with tiny "dentelles" which give it such a rough quality that it was once widely used as sandpaper. Sharks' jaws basically look like they have been coated by this skin.

Two attacks

There it is again! After only the briefest absence, the huge *Dinichthys* has reappeared, its monstrous jaws gaping. It continues to patrol in search of food.

This animal hunts in a very instinctive way; its behavior is not based on any particular strategy or trick (then again, the *Dinichthys* is not very intelligent; but, for that matter, no fish have very high IQs). The movements of this killer are dictated by some automatic mechanisms triggered by simple chemical, acoustic, light, or pressure stimuli. Yet, its brain and its nervous system show distinct growth with respect to preceding aquatic models.

Study of the crania of placoderms of various sizes found in fossil sediments has shown that they had an encephalon that was more developed than that of the modern lamprey. Three pairs of muscles gave the eyes considerable mobility. The inner ear had three semicircular canals for balance, like human beings.

Some placoderms seem to have a structure (the Lorenzini ampoules) destined to contain receptors for electric fields: a kind of "radar" system with which to locate approaching prey (or predators). This system was already present in ostracoderms (the small armored fish that had no mandible) and some scientists surmise that certain protofish used it for defense, producing electric discharges like the electric ray.

There! It has captured an ostracoderm! Circulating incessantly and constantly changing direction like a shark, the *Dinichthys* has managed to take by surprise a small group of ostracoderms near the bottom. Fleeing in panic, one of them chose the wrong direction for flight and ended up crushed between its predator's well-honed fangs.

After swallowing its victim whole, the *Dinichthys* sets off again, mouth agape, in search of food. Its ability as a swimmer allows it to burst into a school of fish and devour the slower ones. Its strength, on the other hand, makes it fearless of even the most terrifying animals. The large scorpions that once ruled these seas, for example, have now been outclassed by placoderms with their jaws and, in particular, their size. In fact, many scientists feel that placoderms, the new predators, were instrumental in the disappearance of the giant marine scorpions.

The *Dinichthys* must have picked up the scent of another prey: it dives down among the rocks. We carefully draw closer to watch it as it tries to squeeze its armored head into a recess to get at a "morsel" hiding there. We can barely see a huge scorpion flattened out in a corner between two rocks in an attempt to escape its assailant's attacks.

But the *Dinichthys* will not be discouraged. Although it can't get at the prey, it continues to press its snout between the rocks. This is a succulent mouthful; the scorpion's carapace is no obstacle for its jaws, which are capable of splitting and grinding it as if it were a shrimp.

All the scorpion has to do, though, is sit still and wait. And the scorpion knows it. If it were to give in to the temptation to flee, it would be finished off in no time.

The *Dinichthys*' attack is unsuccessful this time. After circling around the area for a long time, it finally changes direction and swims away.

The diaspora of fish

The earth's waters are now populated by a growing number of fish of all kinds. Seas, lakes, rivers, and marshes are seething with life. Salt water, brackish water, and fresh water are creating different niches for a variety of life forms.

It is in this period that the ancestors of modern fish are being born and are developing. And it is in this period that there is a kind of evolutionary "diaspora" of the primitive common stock, resulting in a number of great lines destined to have different fates. Some groups, like the ostracoderms and the placoderms, will disappear. Others, like sharks, will survive for hundreds of millions of years. Others will develop bony skeletons and will give rise to the great range of bony fish that we have in our modern rivers, lakes, and seas.

Still others (and these are the ones that interest us) will start to develop increasingly efficient adaptations for leaving the aqueous environment and living on land—initially for brief periods, but then for increasingly longer ones until they finally remain there permanently. This is the part of the story that we are about to tell.

"In 1938, an exceptional discovery amazed zoologists who thought this fish had been extinct for millions of years: a live coelacanth was caught in the Indian Ocean."

Fish with bones

Returning to the seas again after ten million years is quite a surprise—a continuous surprise. The pace of change is incredible. It

"There are curious examples of fish that do not have lungs but that are nevertheless able to leave the water. One of these is the mudskipper."

seems as though evolution has shifted gear—especially if we think back to the exasperating slowness of the first two billion years.

The waters are now slightly opaque because of the microscopic particles suspended in them, but they abound with new colors and new forms of life: algae, sponges, large water worms, ostracoderms, scorpions. But above all there are small fish, unseen until now, which generally resemble modern ones.

They already have a bony skeleton and are now adopting other new characteristics: eyes with a crystalline lens for better sight, taste buds around the mouth, a "lateral line" sensitive to variations in water pressure, and an inner ear containing otoliths. (Otoliths are those tiny calcareous "pebbles," which human beings also have in their inner ear, that shift with the position of the head, signaling the spatial position to the nervous system and setting off muscle reactions to maintain equilibrium.)

It is here, among these ancient fish, that the great revolution—the one that will lead to mammals and therefore human beings—is being prepared.

A number of great lines are splitting. The main one, that of the *actinopterygii* (the first to appear over 400 million years ago) will lead to the seemingly infinite variety of modern fish: there are, in fact, 21,000 known species of bony fish, more than those of all vertebrates combined. This makes it impossible to follow all the ramifications and variations of this immense evolutionary shrub. But two branches are developing in exactly the direction that interests us: the animals of both have rudimentary limbs for walking and rudimentary lungs for breathing. Let's look at the first.

A walking fish

Swimming through the shallow waters, we finally come upon a muddy bank that seems suited to our observations. The water is low and the tiny waves lapping onto the beach cause some turbulence.

Indeed, a strange-looking fish is moving in front of us in a very unusual manner. It has settled onto the ground and is now literally walking around! Its limbs resemble legs, but they are actually two pairs of transformed fins.

Until now, the fins of fish were radial, that is, fan-like, well adapted to work as oars and totally inappropriate for use as props (especially out of water, where gravity is stronger). But the fins of this fish look like crutches able to support the weight of its body. This is an excellent preadaptation for leaving water and moving on dry land.

This fish is the coelacanth and numerous fossil remains have allowed paleontologists to examine its strange "crutches." Unexpectedly, they already have a structure very similar to that of our limbs; that is, they resemble our arms in that they have joints equivalent to our wrists, elbows, and shoulders. These fleshy lobed fins are equipped with a bony inner structure and muscles; they are to all intents and purposes real limbs.

In 1938 an exceptional discovery amazed zoologists who thought that this fish had been extinct for millions of years: a live coelacanth was caught in the Indian Ocean (it was like capturing a dinosaur). Since then, various coelacanths have been caught and studied. Of course, these fish differ slightly from their ancient ancestors (the modern versions live at very great depths), but their structure is still very similar to that of fossil coelacanths and they have provided scientists with an opportunity to observe an important moment in the history of life firsthand.

The coelacanth in front of us moves agilely in the shallow fresh waters of the late Devonian period. But as soon as it notices our presence, it starts to walk away across the muddy bottom—determinedly, but without haste.

Will this be the species to climb onto dry land? The answer is no. The coelacanths and various members of the same group, the coelacanthids, possess this important preadaptation (some of them may even take brief jaunts on the beach); but these animals, which have chosen a substantially aquatic life, still lack a fundamental quality for life on land: lungs suited to long stays out of water. And without an efficient respiratory system, continental life is out of the question.

What animals will be able to solve the breathing problem effectively?

Introducing oxygen into the blood

Breathing on land and underwater basically calls for one thing (both simple and complicated): the introduction of oxygen into the blood.

In does not matter how this oxygen enters the blood (by means of gills, lungs, or whatever); the important thing is that the blood must be oxygenated in some way to satisfy the energy needs of the cells.

There are curious examples of fish that do not have lungs but that are nevertheless able to stay out of water. One of these is the mudskipper.

The mudskipper is a small fish found along the coast of Malaysia and Indonesia. It has developed a surprising adaptation: by taking water into its mouth and sloshing it around, it brings the oxygen contained in the water into contact with the blood vessels lining its mouth so that it can be absorbed. Not much, but just enough to give it a few minutes' autonomy on land.

If the mudskipper then manages to find another puddle or pool from which it can absorb more oxygen (like a driver stopping at a service station to "fill 'er up" along the way), autonomy increases. Some oxygen is also absorbed by the skin when it comes into direct contact with water.

Using these adaptations, the modern mudskipper manages to make raids onto dry land and even climb mangrove trees to catch insects. But of course it cannot live permanently out of water. That would take an adaptation allowing it to absorb oxygen in a far more efficient way.

Such an adaptation—lungs—had almost been completed in the Devonian period.

On shore, we strip off our wetsuits and get ready for the long trek to an area in which we may be able see the real protagonist of the terrestrial landing: a fish endowed not only with limbs, but with very efficient lungs as well. Our destination is distant, though, and lies across a stretch of desert in a place where a lagoon flanks a group of lakes.

Walking through the "old red sandstone"

The landscape we are walking through has been shaped by a series of geological and climatic traumas that took place in the Devonian period. The Devonian was, in fact, a very restless time for our planet. Initially it was marked by violent earthquakes and volcanic eruptions. The movements of the earth's crust heaved parts of the sea floor skyward, depriving them of their watery covering.

These gigantic emersions (and subsequent erosions) gave rise to the great compact sand deposits we are walking on. They are known as old red sandstone (traces of which can still be found in Scotland, for example) and contain enormous amounts of fossils from the Devonian period.

This red plain was once at the bottom of the sea; today it is hot and stifling. There seems to be no air at all. It is as if the plain were still at the bottom of the sea with the water removed. In fact, a cliff that rises in the distance marks the former coastline.

There have also been important climatic changes in the preceding millions of years, with violent precipitations alternating with droughts. The changing of the geography of the oceans has led to the drying up of many areas with internal waters, giving rise to deserts like this and scattered lakes, some of which are enormous. Thus, sea water, fresh water in the lakes and rivers, and the brackish water of lagoons and estuaries have created environments suited to the diversification of fauna that we just witnessed.

The area we are headed for is spotted with close-lying lakes.

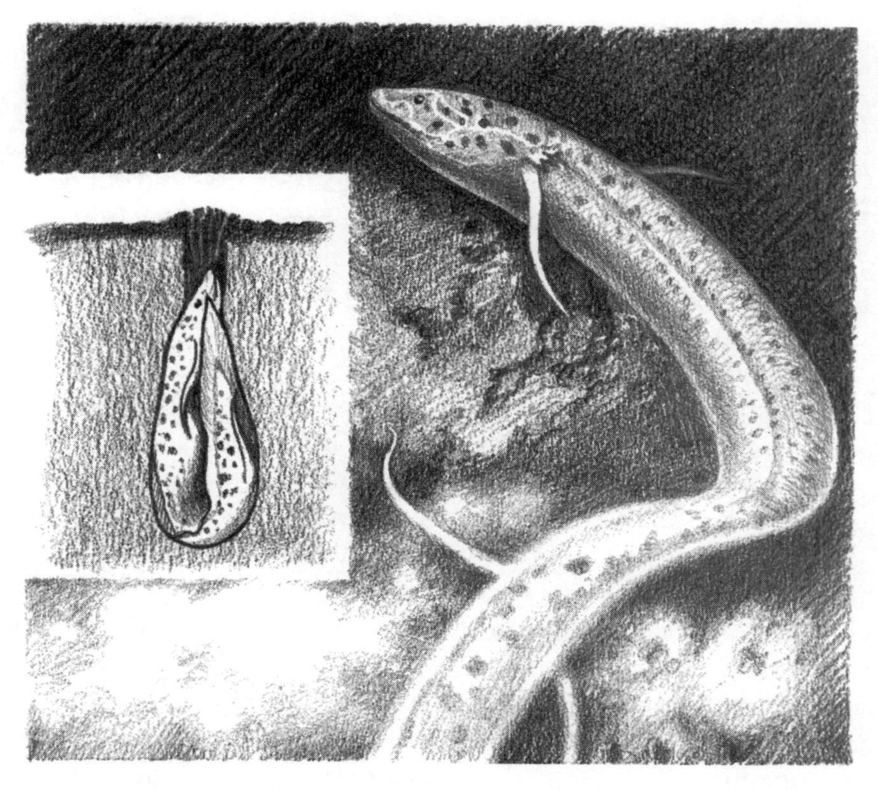

The protopteron: a fish that encapsulates itself in the mud and survives in that condition for months until the rains return. It has gills to breathe when in the water and rudimentary lungs to breathe air directly when in the mud.

This is exactly the kind of environment that can favor gradual adaptations facilitating the transition from water to land; it provides a place (and the time) for development of those fundamental organs that make it possible to absorb oxygen from the air rather than from water: the lungs.

Lungs resembling hernias

As biologist François Jacob observed, evolution is often just a matter of do-it-yourself solutions. That is, it uses old things to solve new problems. New tissues, new organs, and new functions do not necessarily have to be invented from scratch; it would be extremely complicated (and lengthy) for transformations and adaptations to take place in this manner.

Instead, it is much simpler and faster to use things that already exist and modify them for a new function. Sharks provide an excellent example of this principle: the last two branchial arches were transformed into the upper and lower jawbones; around the mouth, the skin with its hard scales was transformed into teeth.

Through constant revision and differentiated developments, evolution offers natural selection a vast range of proposals: some fruitless, some neutral, some successful. Lungs came about the same way.

Modern research shows that lungs originated from a "defect" in the esophagus: that is, from evaginations (hernias) of the esophagus which created pockets or sacs. The inner walls of these sacs, which were lined with blood vessels, were perfect for absorbing the oxygen from the air (in the same way that oxygen is absorbed in the mouth of the modern mudskipper).

Through a series of transformations, this anomaly led in two directions: on the one hand, to lungs and fish able to absorb oxygen directly from the air (pulmonate fish); and on the other, paradoxically, to an important instrument for navigation—the swim bladder. As it happened, some pulmonate fish returned to live in water, but no longer needed the air sac in their esophagus. So it disappeared in some, while in others it was transformed into a kind of hydrostatic balloon, allowing them to swim vertically: another biological invention resulting from evolution's tendency to do-it-yourself and use what is on hand.

How did these primitive lungs work? Fossil remains can do little to help us understand, even though many pulmonate fish have been found in deposits dating back 370 million years. But some of these pulmonates belong to the line of the dipnoids, whose descendants still survive today. Observation of modern dipnoids, which switch between two breathing systems—gills in water and lungs outside of it—suggests that these ancient fish probably had both systems.

Numerous television documentaries have been made about the dipnoids. Some varieties are the protoptera, the *Neoceratodus,* and the *Lepidosiren,* all capable of incredible feats. These fish usually live in water and use their gills. But during the dry season when the ponds dry up, the protopteron, for example, rolls up in the mud, enveloping itself in a waterproof capsule made of mucous to conserve body moisture, and breathes through a tiny hole in the mud. Its lungs basically consist of an evagination with a dense network of blood vessels for the exchange of oxygen and carbon dioxide.

When the rains come and the animal reawakens, it returns to the use of its gills like fish. But if it lives in swampy water which contains little oxygen, it will rise to the surface to breathe with its lungs.

This is a perfect adaptation to both environments, offering a way to absorb oxygen from either, as circumstances require. And it may reflect an ancient need to survive in ponds that periodically dried out (as happened at the end of the Devonian period). Could this mean that some ancestor of the protopteron may have been the Christopher Columbus of the primitive world, the forefather of all terrestrial vertebrates?

Once again, the answer is no. It could have been, but it was not. Fossils remains (found together with those of coelacanths, the four-legged fish) demonstrate that the bones of the cranium could not have been those of the first land amphibians. Furthermore, the fins of these ancient dipnoids were not suited to walking on land. Thus, we will have to look elsewhere for our candidate. But where?

The shrub of evolution, with its many attempts in all directions, shows (as we just saw) that there were different kinds of fish living 370 million years ago: (1) those that could walk on four legs, (2) fish that were taking short jaunts onto land, and (3) those that could absorb oxygen from the air. Were there by chance any that could do all three things?

The star comes on stage

This looks like the right place for our observations: from behind a few shrubs, we have an excellent view of one of the lakes dotting the landscape. The desert is far behind us; here the vegetation around the lakes is lush. This is where preparations are being made for the big event: the landing of our most ancient ancestor.

Unlike the invasion of Normandy, there was, of course, no zero hour for this landing. Things happened gradually with successive adaptations and adjustments. But this 27th of November is a very important date for our planet: after billions of years during which it was no more than a huge field of lava and rock, the earth is now taking on a new appearance. After the arrival of plants and the first minuscule invertebrates, the ancestors of what will be the great protagonists of life on Earth—the vertebrates—are about to conquer the continents.

Vertebrates: the ancestors of the incredible fauna that we admire today ranging from hamsters, bears, and iguanas to camels, snakes, vultures, zebra, turtles, gorillas, toads, horses, chickens, lions, and so forth. All are variations on the same theme and all are part of a score that extends over 370 million years. This was the first note.

Something seems to be moving in the water close to shore.

"The animal stops and looks around and then slowly lumbers onto the beach. It is curious to watch this animal as it moves with difficulty but determination." The *Eusthenopteron* walks onto dry land: an important event in the history of our planet.

A head emerges. It looks like that of a normal fish. But the animal only stops to look around before lumbering slowly onto the beach. It is curious to watch this animal as it stalks forward with difficulty but determination: its body is fishlike and covered in scales, but its four fins are so strong that they double as supports for walking; its mouth gapes rhythmically to take in air. This *Eusthenopteron* looks like an assembly of various parts.

It is anything but small (it measures up to two feet) and its tail sways slightly back and forth as it makes its way up onto land (a gait that characterizes many land animals).

Its fins already have the internal bone sequence of all land vertebrates (1 + 2 + the autopodial bones: i.e., humerus + radius/ulna + hand; or femur + tibia/fibula + foot): a system of successive ramifications. The cranium is made up of strong, compact bones. The choanae (the apertures linking the nasal cavity to the inside of the mouth, typical of all breathing animals) are already present.

In the *Eusthenopteron,* several adaptations that started with ancestors such as the *Osteolepis,* considered the founder of the family line, were completed. The *Eusthenopteron* belongs to the "rhipidists," a group that is distinct from the coelacanths and the dipnoids, but which combines and improves the walking and breathing abilities of both.

With great acumen, someone has pointed out that these three groups adopted quite different strategies to deal with dry land: flight, defense, and attack. In fact, at a certain point, the coelacanths returned to the sea: their adaptation to the shallow waters of lakes proved unsuitable and "flight" to the depths of the sea turned out to be the best way to survive. The dipnoids defended themselves by learning to use lung-type breathing to integrate or substitute their normal gill breathing at difficult times.

Finally, the rhipidists (to which the *Eusthenopteron* belongs), adopted an "attack" strategy, gradually conquering dry land. First they moved from lake to lake or from pond to pond. Then, they became increasingly independent of water. This conquest was difficult, but it had its advantages. The continents were not only sprouting vegetation which turned them into gardens abounding in food of all kinds (both plants and invertebrates), but they also provided shelter from marine predators. Therefore, coming onto dry land meant finding a wealth of available prey, while leaving behind predators, some of which were, as we have seen, very powerful and aggressive.

The first imprint

The *Eusthenopteron* continues its awkward march, holding its body straight on those rigid fins and looking around as if in constant search of something.

This animal is not very evolved with respect to some of the creatures it has left behind in the sea. But it is perfectly adapted to its niche and that is what counts. Throughout the history of life, overspecialized creatures have sooner or later ended up in an evolutionary dead end; the forms that have been open to change are the ones that have adapted best to new environments.

The *Eusthenopteron* slowly disappears into the vegetation, leaving strange tracks behind it in the sand: tracks quite different from those left by the first astronauts walking on the moon, but equally important. They mark the beginning of the conquest of a new planet.

A huge red sun slips behind the horizon. A slight breeze has come up. Before long, it will be dark in this part of the world. Perhaps the *Eusthenopteron* will be able to find a new pond before the sun comes up to warm the world once again. In the meantime, the night belongs to this primitive beast. A night of total silence: no cries, no calls. But a night that is starting imperceptibly to come to life.

One of the most astounding cases of evolutionary convergence is the so-called saber-toothed tiger. From top: *Eusmilus sicarius, Thylacosmilus atrox* (marsupial), *Machairodus megantheron, Smilodon californicus.*

9

Evolution on the March

THE POWER OF DIVERSITY

While the *Eusthenopteron* starts to explore land, we return to our tent to ponder this first part of our journey.

More or less half of our story has now unfolded and the great adventure of land vertebrates is about to begin. In a relatively short time (only a month in our diary), these animals will give rise to amphibians, reptiles, birds, mammals, and finally human beings. Extraordinary diversification from one basic model: four-legged vertebrates.

From looking at our notes one thing immediately becomes clear: life can adapt (and evolution can continue) only if new species continue to develop. It is this constantly changing variety that allows life to overcome obstacles and occupy all available niches.

But why do living forms transform at all? Why does the "right," or rather, "most fit" form always develop for every environment? It may be time to try to answer these questions as they involve an issue that is central to this book: evolutionary processes. Ultimately, it is thanks to these processes that we exist at all and that we are here writing (and you reading) this book.

ONE MIXES, THE OTHER SELECTS

The study of evolution has shown that there are always two great mechanisms at work: one that *mixes*, another that *selects*.

The former acts on the genetic material through accidental recombination of DNA sequences as a result of mutations, replication errors, the action of radiation, or the deletion or insertion of single elements (that is, individual nucleotides) or of entire sequences of segments. This is the way in which new DNA templates and different proteins (i.e., different cell properties and functions) are continually created. The latter acts like a filter on individuals (or populations), eliminating those that stand in the way of survival.

Modern genetics gives much importance to the mutations that are not necessarily favorable or successful—mutations that are simply neutral or that may be associated with an advantageous trait. These pass through the filter and may accumulate, creating what is called a preadaptation in an individual or group. In other words, in the course of evolution, certain new functions do not appear suddenly, as if by magic; they are the result of a gradual transformation (or accumulation) that takes place over time without any specific goal, but which at some point becomes important because it responds to a new environmental request.

Yet what French biologist François Jacob called the "do-it-yourself" quality of evolution also has to be factored in. This term, discussed in the last chapter, means that certain pieces or parts that were useful for a given purpose at one time, may be modified or altered to serve another purpose at a later time (nature provides many such examples).

According to modern biological theory, this is made possible by the enormous variation in living systems and the length of time that it takes these variations to occur: millions or even billions of years. A good analogy is a lottery. There are countless combinations, but also an enormous number of people buying the coupons, trying out all sorts of permutations and variations; most people who play the lottery lose, but in the end there is always someone who accidentally gets all the numbers right.

This is the kind of model living forms have followed in evolution. And this is what has given rise to the wide variety of combinations and recombinations which are then put through the filter of survival. By means of this selection, the environment indirectly shapes certain attributes of living forms—the characteristics needed to adapt to some particular niche in response to the physical, behavioral, reproductive, and nutritional challenges of survival.

The ability to use new sources of food is one of selection's trump cards, especially when it means extracting calories from substances that were indigestible, inaccessible, unnutritious, or perhaps even poisonous to other living forms.

WHICH IS MOST EVOLVED?

It is interesting to observe how the forelimbs of certain animals modified from a common model in response to different functional requirements.

In the following illustration, the adaptations of nine animals (amphibians, reptiles, birds, and mammals) can be compared. All limbs are made up of the same kinds of bones which have, however, adapted by varying in length, shape, and volume; all have been "modeled" by selection.

An interesting question arises at this point: which of these

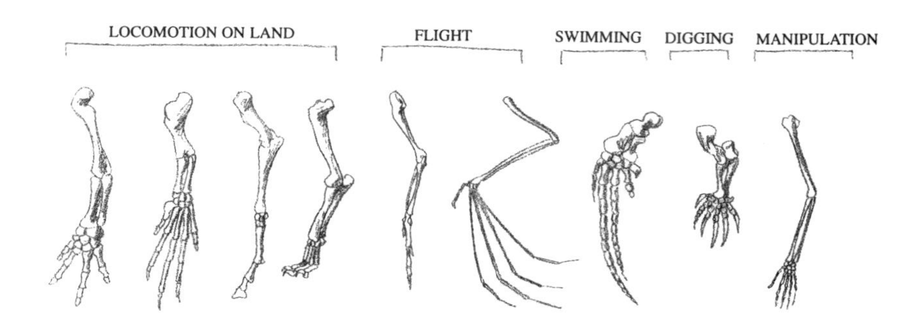

Left to right: The forelimbs of a salamander, a crocodile, a horse, a tiger, a bird, a bat, a cetacean, a mole, and a human being.

limbs is the most evolved? Actually, none. Although it cannot be denied that there has been an ascent up the evolutionary ladder from "inferior" to "superior" organisms, most modern biologists feel that no species is more evolved than another: the important thing is to be well adapted to one's environment through modifications and specializations.

"Assembly" from simple to more complex, which is evident when looking at billions of years of evolution, has already been discussed (and we will have occasion to do so again). But biologists warn us against considering it a process of "perfecting"; it is actually one of "complexification," aimed at responding to environmental requirements.

In fact, evolution does not necessarily tend toward complexity: there are cases in which it tends toward simplification or changes direction. The horse is a typical example: fossil remains attest to a gradual increase in size and a tendency toward reduction of the bone in its hoof at a time when other forms showed the opposite trend.

Basically, modern biology feels that there is no specific "direction" or objective toward which evolution is moving. Rather, living forms develop in all directions through continuing blind attempts which the environment alone selects and eventually models.

THE EVOLUTIONARY TIME SCALE

Evolution obviously works on a very long time scale: it cannot be seen by the "naked eye," so to speak. It is like grass growing or people aging or continents moving: we can only see the changes if we compare certain situations over time.

Evolution is the same. It is at work at all times (even now), but we can perceive it only by making observations over a sufficiently long time span: tens of thousands of years, better yet, millions. Thus we are unable to follow the evolution of a species today from one year to another or even from one decade to another. But the study of past evolution is complicated by other factors as well. We do not have a complete overview of all the species that ever existed. Although an infinite number of species must have existed in the past, we are only familiar with the ones that have been documented in the fossil record, that is, a tiny fraction of the total, since fossilization is a very rare process. In addition, the fact that soft parts do not fossilize makes fossil remains incomplete and fragmentary. For this and other reasons, it is difficult to piece together all parts of the evolutionary puzzle.

Fossils can nevertheless provide us with a general survey of evolution, illustrating the great changes that took place and the important sequences of events. But they can hardly portray the minor intermediate modifications. In order to observe those, we would, in theory, have to have fossils of all the animals that ever lived (or better yet, specimens of their DNA, assuming that we will be able to "read" genetic maps one day and classify genes in a kind of dictionary). But transitional forms usually appear in

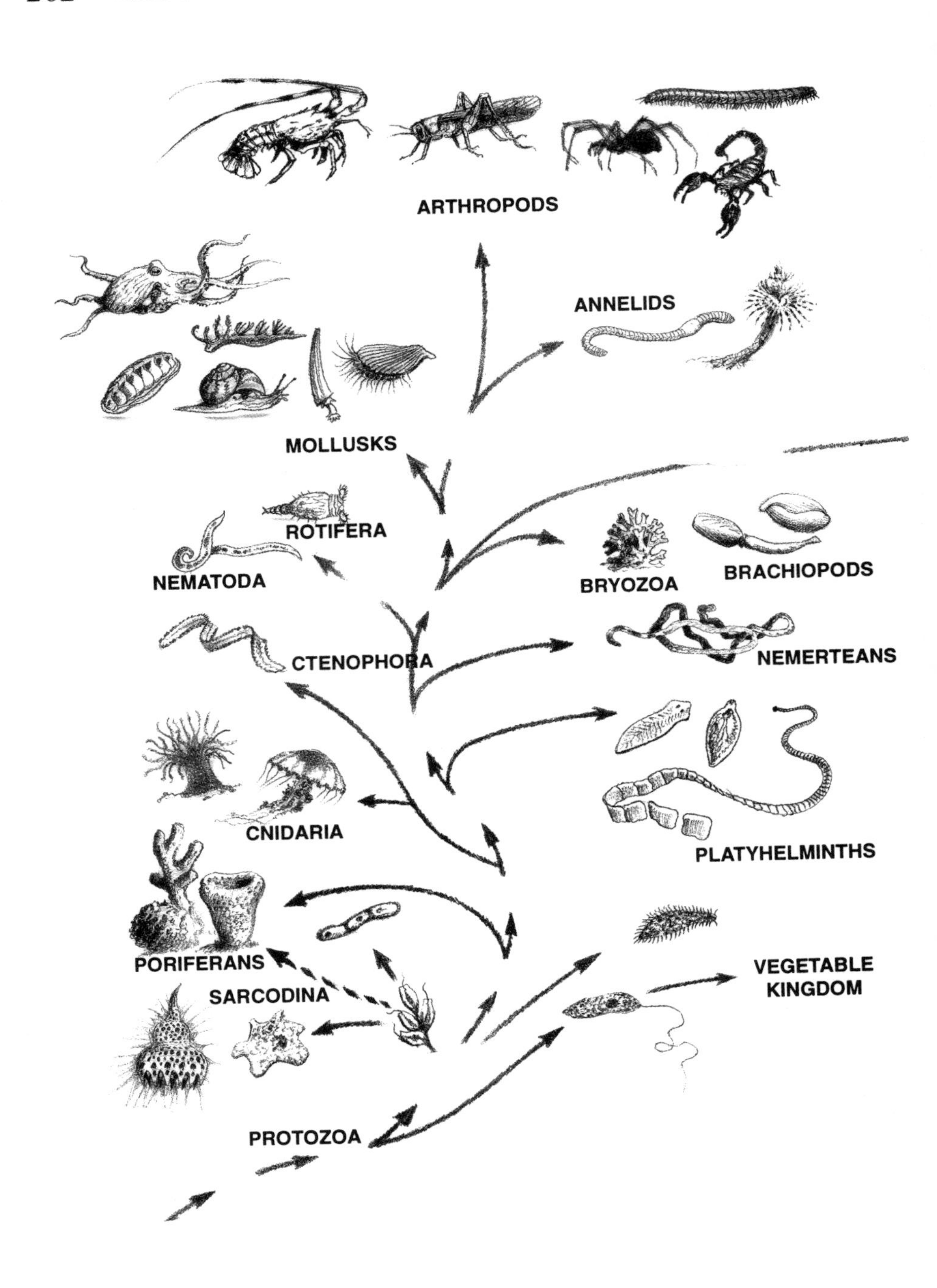
ARTHROPODS
ANNELIDS
MOLLUSKS
ROTIFERA
NEMATODA
BRYOZOA
BRACHIOPODS
CTENOPHORA
NEMERTEANS
CNIDARIA
PLATYHELMINTHS
PORIFERANS
SARCODINA
VEGETABLE
KINGDOM
PROTOZOA

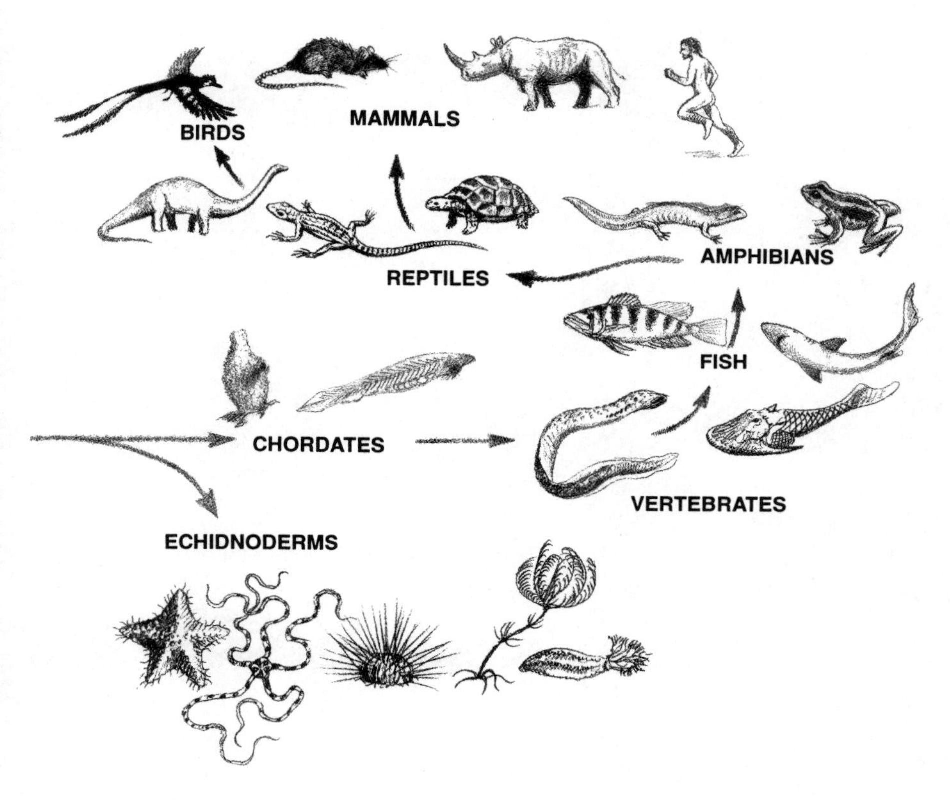

The paths of life: each arrow stands for an evolutionary turning point and the progressive acquisition of increasingly complex characteristics. The appearance of cell aggregates is followed by that of tissues, a mouth, a digestive sac, the mesoderm (which organizes body geometry and structure), a digestive tract, the coelom (the body cavity containing the internal organs), the protostomia and the deuterostomia (two different kinds of embryonic development which lead, respectively, to mollusks, annelid worms and arthropods, and to echinoderms and chordates and, therefore, to all superior animals), the notochord (see p. 134), etc. The arrows make it easier to see which groups "pick up" certain new characteristics and which groups come to a dead end or reroute. This "tree" shows that fish, amphibians, reptiles, birds, and mammals (which seem so different and distant from one another) actually belong to a very homogeneous group: four-legged vertebrates. The origin of the plant kingdom has been indicated by a single sign. (Modified from Storer, Usinger, Stebbins, and Nybakken)

small groups and it is statistically unlikely that an entire series of these transition fossils will ever be found.

In spite of all these difficulties, the evolutionary picture now seems to be quite detailed and clear.

A FEW QUESTIONS

One aspect of the evolutionary process is still puzzling: how can certain mutations take place *at the same time* (or so it seems) to create, through their combined action, a new function or organ?

For example, the development of the eye obviously calls for a large number of transformations (the appearance of the optic nerve, the retina, the eye socket, the crystalline lens, the cornea, and so forth). None of these features could exist without the others. How can they all have developed together? And how can they have developed harmoniously, fitting together perfectly?

The same is true of many other features. Each anatomical, functional, or organic transformation (whether large or small) calls for a series of adaptations that often involves neighboring and sometimes even distant areas and that has to take place in a synchronized way. It is difficult to believe that all the adaptations that occurred in the past took place by chance (even if chance has played an important role in evolution).

In order to answer this question, a better understanding must be gained of the embryonic development of the individual. In fact, separating a DNA mutation from the corresponding transformation at the individual level is embryonic development. Development of the embryo is the fundamental moment in which a genetic mutation is *processed* or *interpreted,* one might almost say. This interpretation activates or inhibits the action of other genes and special regulatory mechanisms.

Much still remains to be learned in this field. Today, embryology represents one of the most stimulating horizons of modern research, but also one of the most difficult fields to explore, given the technical difficulties involved.

Researchers are, however, beginning to understand a number of things. And a number of others are being hypothesized. So let's take a closer look at this crucial phase of transition (and linkage) between genes and their individual expression. (The following pages will inevitably contain a few splinters of glass. But we feel that the explanation is generally rather comprehensible.)

THE SECRETS OF THE EMBRYO

The beginning of the story is well known: a sperm fertilizes an egg. Each brings with it one half of the chromosomes needed to produce a complete individual.

The penetration of the sperm triggers cell division. At a certain point, the cells resulting from these divisions start to differentiate: some become skin cells, others bone cells, muscle cells, nerve cells, and so on. After a certain number of cell divisions (fifty-six in the human being) the individual is complete, formed by billions and billions of cells linked to one another in a single well-organized system. Understanding how this takes place is one of the key concerns of embryology.

Let us, therefore, return to our fertilized egg and try to get a closer look at it—an inside look, to be exact. We are going to enter it.

After having crossed the cytoplasm, we penetrate the nucleus, which contains a terrible tangle known as *chromatin*. Chromatin is made up of various things, but the main component is DNA, the famous double-helical molecular structure that acts as a template for all cellular material. In the early stages of division, the DNA is "blocked," so to speak; that is, the cells can only duplicate themselves. In fact, during these initial stages, all cells are the same. But as duplication continues, things change and differentiated cells begin to appear.

This process of differentiation evidently indicates that something has occurred in the nucleus, the place where the commands for construction of the cell originate. But what is it exactly that happens?

The current hypothesis is that "activation factors," that is, chemical compounds (often proteins) that can activate or repress the expression of certain genes, enter into action. The areas of DNA thus activated could send out new information and start to have it translated by the cells.

In other words, DNA is like a keyboard with some keys that do not produce sound at the beginning. Then, as the activation factors gradually bring new keys into play, the sounds they produce combine to form melodies and harmonies, or rather, forms and structures.

But another fundamental question is why are different genes activated in different cells? And why does it all seem to correspond to some kind of logic, creating forms and structures that fall into the right place instead of appearing randomly?

Once again, we have no more than hypotheses to answer these questions. It is believed that each cell is equipped with a clock and a map, so to speak, meaning that it changes in time and in space and that there is a very special mechanism behind it.

The succession of cell duplications probably produces (with the passing of *time*) biochemical alterations within the cell, causing "maturation." This maturation is in turn influenced by the context in which each cell finds itself, that is, the kind of molecules with which it comes into contact. Basically, it depends on the proximity (or lack of it) of certain other cells. (In a certain sense, the same thing happens in human beings, especially with respect to cultural development: with time, each person undergoes a maturation that is influenced by the ambience in which he or she lives and works, and the kind of people he or she enters into contact with.)

For example, it has been found that if one hundred identical cells are grown in a row away from a single point of reference, the influence of the point of reference and certain individual properties of the cells decrease as the distance from that point increases, while the influence of casually encountered cells increases.

This process of reciprocal induction, activation, repression, and inhibition is fascinating. For example, the future retina of the eye can be seen growing at the end of two rows of cells (the optic nerves in formation). When it comes to the surface, it induces the skin to fold back to form the socket and then the crystalline lens, which in turn induces the adjacent tissues to form a transparent membrane, the cornea, and so forth.

Cell constructions have the ability to model themselves through successive adjustments, taking account of what is going on around them and adapting themselves flexibly to new chemical and genetic signals. During construction, the various parts influence each other reciprocally.

Technological evolution is similar. The first model of the automobile, for example, did not merely consist of a chassis, to which wheels, a steering wheel, brakes, and other parts were added at a later date. There never were such intermediate stages and it would be futile to look for them. In the same way, the eye did not start with a retina to which the eye socket, the crystalline lens, the cornea, and so on were subsequently added by chance.

Both technological and biological evolution always start out with a simple model (in the case of sight, probably with simple light-sensitive cells or tissue), but only functioning (and environmentally compatible) "wholes" can entertain hopes of entering into the system of life and reproduction.

Scientists distinguish three layers in the development of an organism: the first layer (the *ectoderm*) produces the more external parts of the body, such as the skin, the sense organs, and the nervous system; the second layer (the *mesoderm*) organizes the geometry of the body and provides the supporting structures (bones, muscles, connective tissue); the third layer (the *endoderm*) is responsible for the internal digestive and respiratory organs (such as the stomach, liver, intestines, and lungs)

In order to understand the functions of these various strata, attempts have been made in the laboratory to invert their positions.

In chickens, body feathers and the scaly skin of the legs are usually produced by the first layer, on instruction from the second. Inverting the positions of the first and second layers can produce a chicken with feathers on its legs and scaly skin on its body. And if the orientation of the first layer is inverted, the chicken may have all its feathers the wrong way around!

In the first embryonic stage, the entire construction of the individual can be changed. For example, if a piece of tissue that would normally develop into an eye is removed from an amphibian embryo and transplanted into the tail region of another embryo, it will no longer develop into an eye, but rather into a kidney duct or a part of some other organ in the area.

At a more advanced stage, this extreme adaptability decreases (just as the cultural adaptability of an individual decreases with age) and the transplanted tissue will develop into an eye no matter where it is transplanted—even on a thigh.

THE TYPEWRITER AND THE CHESS GAME

There is no doubt that embryology is one of the research fields that will provide us with a greater understanding of evolution in the future. But current knowledge suggests that some kind of gene "modulation" takes place between the reading of DNA and its expression in the individual, and that this modulation is influenced by regulatory mechanisms and the context in which the cells develop.

In other words, a mutation in the DNA keyboard is not like substituting a letter on the keyboard of a typewriter (which results in that one letter being different every time). It is more like replacing a piece of a chess game; *all* relations are changed, not only those between the piece that has been changed and the others, but indirectly among all pieces on the board.

This analogy facilitates understanding of certain "chains" of changes that would be difficult to explain simply as the result of a series of independent (and synchronous) mutations. It also makes the evolutionary process more comprehensible, as it highlights the flexibility and malleability of the context in which genes operate.

To give only one example: the extraordinary variety of radiolarians seen in chapter 3 depends only on minor genetic variations which have, however, produced quite different developments and adaptations in the final configurations. (It would be futile to look for intermediate forms, that is, those "missing links" between one radiolarian and another).

Analogously, the bone tissue of any mammal (from the rhinoceros to the dog) starts out from similar genes that are then "modulated" by the context to build different skeletons. The same is true of the genes producing retractile proteins: they are exactly the same in the amoeba and the human being—only their "expression" is different.

What is important for the individual is that the genetic *mix*, that is, the team of genes, responds well to the environment in which it is called upon to express itself and to construct a functioning organism. Mutations are generally unfavorable; indeed, they may be fatal. But in the lottery of change, "wrong" or losing combinations come up all the time. For example, it is believed that the frequency of miscarriages in human beings is very high (many take place before the woman even realizes that she is pregnant). In addition, many other individuals die shortly after birth or during infancy.

Of course, the teamwork of the genes continues to be decisive throughout an individual's existence, creating predispositions to diseases that can eliminate that individual before he or she has reproduced. In fact, although we tend to forget it, natural selection was very strong among human beings in the past, allowing for the survival and reproduction of only those who overcame the test of very restrictive screening.

MORE IMPORTANT THAN THE INDIVIDUAL AND THE SPECIES

As we have said before, it is through this mechanism that all species have been indirectly modeled (in the most diverse ways) by the challenges of life. In fact, although genetic sequences are produced by accidental mutation, they are selected by the rigid requirements of the environment. As French biologist Jacques Monod put it, chance constantly has to come to terms with necessity.

There are, of course, various levels to this endless game of elimination. We tend to give great importance to the individual level, but from an evolutionary point of view, the individual is of little consequence: the species is what counts. There can be no doubt that the survival of the individual is secondary to that of the species. But even the species is not that important from an evolutionary point of view. In fact, it seems that 99.99 percent of all species have disappeared through the course of evolution (and we did not even notice). What really counts is the continuation of life.

It makes no difference whether one or a thousand species become extinct (because of environmental or climatic catastrophes); what is important is that there are always others capable of surviving and reproducing. It's like in a relay race.

The following chapters will give an idea of the many catastrophes and mass extinctions that have accompanied the history of life on Earth. Yet, they have in no way interrupted evolution; on the contrary, they have in some ways spurred it. Like the death of an individual, the death of a species opens up new spaces, allowing for turnover and new births. In fact, the great extinctions of the past led to the advent of other animals—mammals (and eventually the appearance of the human race).

In some ways, extinctions (like cultural crises) should be seen as extraordinary opportunities for unblocking previous situations and accelerating evolution, often by taking directions that were formerly inaccessible.

EVOLUTIONARY CONVERGENCES

Before concluding this brief chapter on the mechanisms of evolution, one final consideration must be made of so-called *evolutionary convergences.*

Curiously, natural history is studded with animals that resemble each other or have very similar characteristics even though they derive from quite different evolutionary lines. It is as if their evolutionary courses underwent similar influences, giving rise to parallel shapes and behaviors.

Do you recall the fish (the protopteron) that lives encapsulated in the mud when the ponds dry up during the dry season? Well, an amphibian (which will be discussed in the following chapter) has developed exactly the same behavior in a totally independent manner.

Just as two unacquainted mathematicians may solve the same problem in an identical way, certain animal species end up finding similar solutions to analogous problems of survival. For example, long legs are advantageous for jumping on sand and, indeed, the legs of certain mice in Egypt and of others that live in New Mexico have been modeled in the same way; both animals also have long tails to help them "take off" when pursued.

Strangely enough, their predators are also the result of evolutionary convergence: the fennec of the African deserts and the American pygmy fox both have enormous ears as well as similar dentition and hunting and living habits. Selection has affected them in the same way. Likewise, the African leopard and the South American jaguar have both developed the same kind of spotted coat; in addition, there are black (melanic) versions of both.

"Flying squirrels" are little animals which, at different times and on different continents, have developed a patagium (a membranous skin linking their superior and inferior limbs), allowing them to leap from one tree to another.

The most surprising thing is that certain marsupials originating in Australia or Tasmania have produced perfect convergences with placental mammals. A famous example is the thylacine—such an impeccable marsupial version of the wolf that it is known as the Tasmanian wolf.

Convergences also occurred in prehistoric times. For example, a reptile called the mosasaur and a mammal called the *Basilosaurus* developed exactly the same adaptation to the marine environment, taking the shape of a sea serpent. Indeed, many of the animals that adapted to the marine environment (ichthyosaurs, dolphins, and cetaceans in general) assumed the exterior shape and many of the abilities of fish.

But there were also convergences between even more distant animals: both the glyptodont (a mammal) and the ankylosaur (a dinosaur) had a huge armored dome on their backs and a club-like tail.

The most surprising example may be the so-called saber-toothed tiger. Almost identical-looking animals can be found among the marsupials and the placentals: perfect convergences occurred at different times between animals coming from different and sometimes very distant evolutionary branches (*Omoth-*

erium, Smilodon, Thylacosmilus, Eusmilus, Machairodus, etc.). All of these animals had not only extremely long and sharp fangs, but also longer front legs, a short nose, and similar behavior.

The list could go on. But apart from the curiosity of these parallel transformations, the phenomenon of evolutionary convergence is extremely interesting in that it allows us to understand how nature operates in shaping its models by selection.

Just as a cold environment requires all animals living there to have a protective fatty layer and an aquatic environment calls for a hydrodynamic shape, various other requirements result in parallel adaptations, even in environments lying hundreds of thousands of miles apart. A typical example is that forest animals are generally smaller than savannah animals (this holds true even in the human race: the pygmies live in the forest, while the Masai and the Watussi live on the savannah). And when requirements become even more demanding and adaptations require very specific and specialized solutions, these convergences may become particularly accentuated and in some cases spectacular.

Then again, a multitude of environmental factors contribute to modeling living forms through natural selection: climate, vegetation, temperature, humidity, and soon. In addition, the living forms themselves influence each other, giving rise at times to curious parallel evolutions. It is enough to think of flowers and pollinating insects (some orchids, for example, in order to attract bumble bees, have taken on the shape and colors of female bumble bees). Prey and predators also model each other, refining their respective abilities for hunting and flight through selection.

We will return to this matter in the second part of our journey. Now it is time to turn out the camp light and wait for a new day to break so that we can follow the enthralling adventure of vertebrates on dry land and observe their evolutionary progress.

The first step involves animals that live halfway between water and land: the amphibians.

"There it is! An *Ichthyostega*. No, two of them. Large parts of their bodies emerge as they come to the surface in the brackish waters of the lagoon."

10

The Earth Becomes Populated: Amphibians and Reptiles

NOVEMBER 29TH (355 MILLION YEARS AGO)

In the hot, steamy swamps

The setting around us seems unreal. The vast swampy area through which we are walking is covered by a dense forest of giant horsetails. The air is hot and humid, almost like a bathroom after a shower. At this time in the late Devonian, the first real forests heralding the arrival of the Carboniferous period are starting to appear.

Enormous changes have taken place since the first mosses battled to conquer the continents: vegetation now includes horsetails, ferns (some huge), and towering trees with scaly bark like the skin of a fish or reptile (*Sigillaria*).

Finally we reach the calm waters of the lagoon lying just beyond the swamps. Warm and only slightly oxygenated, these waters are not very suited to fish life. And indeed, there aren't many fish around: bony fish, in particular, have all gone back to the sea. Amphibians now dominate the environment, both above and below the water's surface.

This is the place where we watched the *Eusthenopteron* lumber onto the bank on its fins-turned-legs millions of years ago. For a long time, the coastline between the sea and dry land was its kingdom; now other protagonists have displaced it.

Structurally, the *Eusthenopteron* was a marine animal with accentuated fish-like characteristics: it had fins for swimming, a scaly skin, and a "lateral line" (a sense organ) like fish. It was capable of moving through shallow waters to get to the inland ponds to capture fish without too much competition from rivals. Its sharp teeth are an indication of the fact that it ate fish, not insects or spiders. In some ways it was like a land eel, with its strong limbs for walking and, above all, its rudimentary lungs for absorbing oxygen from the atmosphere.

But it may not have ventured onto dry land to conquer it at all. Paradoxically, it may have done so only to be able to continue to live in the water, and to be able to reach pools with abundant food at a time when fresh water was starting to be depleted of food. Certainly, this was a very important asset for its survival, but an even more important preadaptation for its descendants. The *Eusthenopteron* and other members of its suborder, the so-called rhipidists were, in fact, the forefathers of the amphibians and, in particular, the animal that we are now searching for in this hot, muggy place—the animal thought to be the link between these two groups: the *Ichthyostega*.

AN EARLY AMPHIBIOUS MODEL: THE *ICHTHYOSTEGA*

There it is! An *Ichthyostega*. No, two of them. Large parts of their bodies emerge as they come to the surface in the brackish waters of the lagoon. Then they remain motionless. This area is rich in food and the *Ichthyostega* are ready to take advantage of every opportunity, both in water and on dry land.

We inch forward carefully trying not to scare them (but they look perfectly at ease). Their resemblance to the *Eusthenopteron* is remarkable, as though a sculptor had created an amphibious version of the same animal. *Ichthyostega* and *Eusthenopteron* have similar dimensions—about three feet in length—and the same shape. But a few things have changed; these newer models have some very important additional features.

The legs are still apparently designed for swimming (just as the flat finned tail shows a strong adaptation to water), but the foot structure is already similar to that of land vertebrates, with joints and bone structures that prefigure ours. The spinal cord is thick and strong to support the weight of the body on land. The teeth still seem to be adapted to fish-eating, but they already have the internal labyrinth-type structure typical of amphibians (and of the *Eusthenopteron*, their ancestor). Like all amphibians, they have no neck and the limbs are linked to the spinal cord by means of connective tissue and ligaments. A structure especially suited to walking on land.

One of the *Ichthyostega* has slipped back underwater. It evidently did not appreciate our presence or had more important things to do. The other is still there, however, immobile; only one eye moves ever so slightly. It is an eye that is adapting to air, an eye that is developing tear glands and eyelids to keep it wet and clean, respectively.

These animals have also adapted their sense of hearing to the new environment. Their middle ear now has an amplification system, a kind of resonance chamber with a bony rod. The rod will develop into the famous stirrup bone or stapes, one of the three bones that we have in our middle ear (the other two, the hammer and the anvil, will appear later through the migration and transformation of two bones of the mandible).

For the first time, the tongue, which is to become the primary taste organ, also seems well defined.

The *Ichthyostega* now lazily leaves the water and saunters onto a mound of wet soil. Its front legs move in a very articulated manner. It looks like something halfway between a crocodile and a tadpole: a large tadpole with legs. It no longer has scales like the *Eusthenopteron*, but a real epidermis, flexible and above all endowed with a mucous layer to preserve humidity.

The epidermis also has pores, allowing the *Ichthyostega* to breathe more easily. In fact, even if its lungs are more efficient than those of the *Eusthenopteron*, breathing through the skin will remain a primary requirement of all amphibians, for whom semiaquatic life is a way of avoiding dehydration.

The *Ichthyostega* ambles forward to a little shady bay. Like a bather, it edges into the water, generating a series of tiny concentric ripples. It has strong plates on its head and, although it is not visible, a "third eye" in the center of its cranium: a structure for pineal vision, destined to disappear gradually in its descendants.

The *Ichthyostega* is not the only representative of this group of primitive amphibians; there are others (the *Acanthostega* is even more ancient). But it is from this shrub, which still has to be reconstructed in all its ramifications, that the real amphibians and later the reptiles will emerge.

Slowly, the animal's head sinks into the water and disappears like a submarine.

DECEMBER 3RD (310 MILLION YEARS AGO)

The great forests of the Carboniferous period

Day has just dawned and the sun has not yet had the time to dispel the mist lying on the land. The humidity is so great that steam seems to be rising from the underbrush. The air is rank with the smell of rotting wood: huge tree trunks, blackened by the rain, are half hidden by the vegetation underfoot.

After crossing a swampy area full of ferns (and huge insects), we enter an immense forest with trees so tall that they form a kind of cathedral far above our heads. These are *lepidodendrons* ("scaly trees"): they can reach a height of 120 feet and a diameter at their base of five feet. Their size and grandeur bring to mind the great colonnade of the Temple at Karnak. The bark of these trees is quite unlike that of trees today: it vaguely resembles the surface of a pineapple, with spiraling scars.

But what is most striking about the forest is the silence: all that can be heard is the gentle buzzing of insects as they fly by. Any other sounds are being produced by human beings—us.

These trees—lepidodendrons, sigillarians, and calamites (all giants over ninety feet tall, i.e., as tall as a ten-story building)—that will give rise to the decaying wood deposits that will eventually be converted into coal. By falling into muddy swamps where air is scarce, the wood is transformed by bacterial action, which withdraws the nitrogen and oxygen and leaves a carbon concentrate that will later turn into coal (a process that is still underway in swamps today).

Of course, all evolutionary phenomena must be considered over an extremely long time frame. The Carboniferous period (that is the period of the great forests) lasted 70 million years, from 350 to 280 million years ago. Throughout that time, dead trees accumulated, creating enormously thick layers with the weight of each compressing and compacting the one below it.

Today, these ancient coal seams can be seen in mining tunnels. The richest seams are the oldest ones. The process is similar to the one that produced oil (in more recent times) starting with microorganisms, in particular, microscopic algae, but also plants trapped in the muddy bottoms of swamps, deltas, lakes, and oceans. In the case of oil, the original matter was also processed by bacteria, which created a variety of hydrogen and carbon compounds (hydrocarbons) that then impregnated the sediments.

A snake! For a moment we thought that long black thing we were about to step on was a snake. But there are no snakes in the Carboniferous period. It is far too early. It will still be millions of years before they appear (when salamanders finally lose their legs). So what was that strange animal that we caught a glance of in the foliage?

It now marches out into the open: its gait leaves no room for doubt, it is a giant millipede. An enormous millipede, which may have grown to defend itself from scorpion attacks. Scorpions still abound in this period (practically identical to modern scorpions), along with spiders and insects.

Insects, the most recent arrivals, have already occupied many of their typical niches and have proved particularly ingenious in exploiting new spaces. In fact, they have colonized an environment which no animal was able to conquer till now: the air. This is the first period in which living creatures manage to rise from the ground and fly.

The conquest of the air

In the course of evolution, many species will independently develop the ability to fly (in particular, pterodactyls, birds, and bats), but this is the first time that living creatures manage to overcome the force of gravity on land, rise into the air, and move around without touching down.

It is a real marvel if you think of it. We are so used to seeing insects fly today, that we don't even think of the difficulty of it all, it seems perfectly natural. But inventing flight was one of the greatest achievements of all time (not even Leonardo da Vinci could manage it). So, how did these first flying forms succeed?

Actually, not much is known about the development of wings. There are a few hypotheses, however. Some suggest that insects' wings were converted from the ancient gills of aquatic arthropods, that is, structures that are obsolete out of water. This would be in keeping with the do-it-yourself nature of evolution, which tends to take something old that is no longer useful and, under changed conditions, transform it into something new and more functional.

Flight has, of course, been very successful everywhere in nature. It has been adopted by a myriad of species with a wide range of adjustments and variations.

Here in the Carboniferous period, for example, there is an insect with six wings, the *Stenodictya*. Insects usually have six legs and four wings (at times only two functional ones), but this one has six legs and six wings; the first two are very short and rigid and seem to serve as stabilizers.

The dragonfly has done something even more difficult: it has invented hovering, that is, static flight. It's basically a helicopter. We saw a lot of dragonflies while we were crossing the swamps on our way here and were particularly impressed by one species, the *Meganeura,* an enormous dragonfly with a wingspan of about two and a half feet. And yet, even this flying machine is as light as the wind, capable of flitting, circling, and dancing in the air.

This galaxy of insects, spiders, scorpions, millipedes, and snails has been able to evolve in solitude for a long time, without the cumbersome presence of vertebrates. And the same is

The *Meganeura,* an enormous dragonfly with a wingspan of about two and a half feet. Box: the sparrow gives an idea of the size of this ancient insect.

true of plants. The first plants "landed" 420 million years ago; the first vertebrates only arrived 370 million years ago. This means that life on dry land evolved for 50 million years as if in a secluded "terrarium."

What would have happened if plants and invertebrates had remained alone on dry land? Would more evolved forms of life have emerged? Most scientists do not think so, since the most ancient inhabitants have remained almost unchanged and are still almost identical to the way they were hundreds of millions of years ago.

Instead, the arrival of the "larger animals" changed everything. The arrival of the amphibians accelerated the process of diversification that was to lead to reptiles and later to mammals, in a crescendo of shapes and evolutionary "inventions." And it is in the extraordinary forests of the Carboniferous period, in these hot and humid niches, that the terrestrial adventure of the vertebrates began.

DECEMBER 5TH (280 MILLION YEARS AGO)

An amphibian wearing Napoleon's hat

For the last two days of our calendar (over 20 million years in the lower Permian period), we have been traveling the continents, discovering the wide range of amphibians now populating them.

Starting with the ancestors of the *Ichthyostega* family, evolution seems to have gone wild creating the most unusual and surprising shapes. (And it will continue to do so. A very surprising amphibian, for example, the *Mastodontosaurus,* will emerge in the upper Triassic, with a total body length of fifteen feet and the head alone measuring three feet.) These amphibians are quite different from the few amphibians such as frogs, toads, newts, and salamanders that managed to survive extinctions and various catastrophes until modern times.

Our notebook contains some examples of this diversity. We have seen amphibians with huge plates of armor on their backs, serpent-like amphibians without legs (a most astounding evolutionary convergence that makes them identical to snakes), amphibians with great bony crests, amphibians that look like fish, others that look like reptiles, large amphibians, small amphibians. . . . A line of amphibians has even developed in the oceans with smaller legs and a strong tail (a little like saltwater crocodiles).

The swampy forest environment provides most of these amphibians with an almost inexhaustible source of food consisting of insects and other invertebrates. But many large amphibians require something just a little more filling, and so they have started to hunt smaller amphibians. Others, finally, have rediscovered ancient hunting grounds: the marine world.

We have donned our fins and goggles to "patrol" the lagoon and see what is happening below the surface.

For some time now, we have been watching a *Diplocaulus* below us, a strange amphibian with a triangular head which

"It looks like a large salamander that has swallowed a boomerang. The *Diplocaulus,* about one yard long, is flattened out on the bottom of the lagoon. . . ."

makes it resemble a large salamander wearing Napoleon's hat. Flattened out on the bottom of the lagoon, the three-foot- long animal occasionally stirs up the mud when it moves its head. In fact, its "hat" is rigid, formed by a bony structure shaped like a boomerang. But its limbs are small and weak: they do not look as though they would be much good for walking on dry land. Indeed, the *Diplocaulus* is in its element at the muddy bottom of lakes and lagoons where it is proficient at digging up the mollusks that abound there.

The *Diplocaulus* is obviously not what we were looking for, so we will leave it to its own devices. What we are looking for is the *Eryops.*

The shark attack

The *Eryops* is a powerful predator that ambushes its prey. It should be lurking around here somewhere. Almost six feet long,

The *Eryops* is a powerful predator that can reach lengths of up to six feet. The *Seymouria* resembles a miniature dinosaur or a small crocodile about four and a half feet long. The *Cacops* looks like a real reptile, but it is an armored amphibian no more than one and a half feet long.

it resembles a cross between a crocodile and a pig, or perhaps a crocodile that has had its head and its tail jammed into its body.

This animal is difficult to spot because, like the crocodile, it is perfectly camouflaged when it lies motionless. It can easily be taken for an submerged log or merely a shadow.

We saw one crawl across a muddy bank and enter the water not long ago. It should be hidden somewhere around here, but we cannot seem to locate it.

These waters are full of amphibians and a kind of small shark, the *Orthacanthus,* which are among its favorite prey (the remains of these small sharks have been found in the digestive tracts of numerous fossil findings of *Eryops*).

An *Orthacanthus* is swimming by right now. It has already circled around twice in search of prey. The *Eryops* must surely have seen it; it may even have seen it before we did. That may be why it slithered into the water.

Suddenly a burst of mud rises from the bottom as the *Eryops* shoots out like a Polaris missile. All we can see as it attacks the *Orthacanthus* is its huge gaping mouth framed by razor-sharp teeth. Little else is visible in the murky water. But the small shark seems to have escaped its predator. A few sharp flicks of its tail, and it is gone. No luck this time for the *Eryops.* It will have to start hunting all over again.

The cloud of mud has diffused, blackening the waters and sending all forms of life speeding off in various directions. We do the same.

The transition toward reptiles

At this time at the end of the Carboniferous period and the beginning of the Permian, amphibians are not the only animals to live on dry land: the animals that will dominate land, water, and air environments for the next 200 million years and more have already appeared and are now developing: reptiles.

During our most recent explorations, we encountered strange amphibians that resemble snakes in every way. Two examples are the *Ophideropton* and the *Phlegethontia*; if we were to run across these creatures today, we would certainly take them for reptiles.

Other amphibians, such as *Seymouria*, are almost identical to lizards. A *Seymouria* actually looks like a miniature dinosaur, a kind of crocodile with long legs. For some time it was considered the missing link between amphibians and reptiles, but then it became obvious that it was only an evolutionary convergence, one of the many that have taken place in the course of evolution. It is an apparently similar, but substantially different creature: not a relative, much less an ancestor.

The transition from amphibian to reptile is still unclear: there is an entire shrub of shapes with a multitude of branches going in different directions; among these are animals that look like reptiles but are still amphibians (like the *Diadectes,* a large lizard almost nine feet long that resembles an iguana).

The *Hylonomus,* the most ancient complete reptile ever found, dates back approximately 300 million years. It was eight inches long.

One animal that does emerge from this jumble of shapes, however, and that paleontologists think may be the missing link between the two worlds is the *Hylonomus*. Its remains have been found in silt-filled tree trunks lying in geological strata dating back to the Carboniferous period. The animal grew to about twelve inches in length and resembled a lizard with a long tail and slender, flexible fingers.

Studies seem to point to the *Hylonomus* as the first completely terrestrial vertebrate, that is, an animal no longer dependent on water and completely adapted to life on dry land. This means that its lungs were improved to the point that it no longer required supplementary breathing through the skin. But above all, it no longer required water for reproduction. This was a real revolution for vertebrates, a revolution similar to the one carried out by plants and invertebrates when they transferred to dry land.

Of the three prerequisites for the transfer from water to dry land (respiration, support structures, and reproduction), the amphibians have satisfied only the first two. The third, reproduction, continues to be carried out like fish. Amphibians return to water for reproduction, depositing their eggs and raising their young (which are endowed with gills) in a completely aquatic environment. Only at a certain point in their lives do the gills disappear. Then and only then are the young able to abandon water and walk and breathe on dry land.

This system is very efficient, but it limits their freedom of movement; water must always be close by. For reproduction (and also to prevent dehydration), amphibians must stay in the vicinity of swamps, lakes, or rivers; inland pools are even better, as they are certainly more tranquil and harbor fewer predators (also keen on their eggs).

How did this transition to total independence from water come about?

The birth of the amniotic egg

Observation of some small salamanders shows that they deposit their eggs both in and outside of water, although always in a humid environment. Some ancient amphibians may have started in the same way, depositing eggs in places less accessible to predators (dry land, in this sense, was safer at that time). And this habit may have led to what was to represent the final transition for vertebrates from water to dry land: the amniotic egg—basically, a kind of "space capsule" that allows the young to develop even in hostile environments.

The basic idea behind this kind of egg (which has a hard shell and contains liquid and nutritional substances inside it) is to close a bit of marsh in a container so that the "tadpole" can continue to develop in an aquatic environment which has, however, been transferred to dry land.

There are many theories about the origin of the egg. It is rather widely accepted, however, that many experiments were carried out by one or more groups of individuals and that the kind of amniotic egg found in nature today is the result of the various attempts made toward an increasingly efficient system.

Inside the egg, the "tadpole" can feed, eliminate its waste, and exchange gases (carbon dioxide and oxygen) with the outside thanks to the porosity of the shell. It can continue to develop its gills (something that will continue in reptiles and even, partially, in mammals, including humans), and finally be ready to enter the world of respiration when its development is complete. Thus, the metamorphoses of the tadpole take place directly inside the egg, which must be seen as a kind of mobile home.

The originality of this invention is that it does not call for the embryo to undergo profound physiological transformations: it continues to develop in an aqueous environment as before. It is the egg that has changed, becoming larger and more rigid. And this has favored its separation from its marine nursery.

In the traditional development from tadpole to amphibian, the various stages were (and are) perfectly visible: in some ways, the fertilized egg gradually goes through all the stages of the amphioxus, the fish, the *Eusthenopteron*, the *Ichthyostega*, up to the amphibian. In other words, the various stages of development present characteristics and behaviors very similar to those of each of these creatures, as if the course of the species' evolution were "recapitulated" during the development of each individual.

In the rigid egg, these different stages all take place inside a shell that will only open at birth to reveal a fully developed individual. But inside the egg, these stages still take place and mark, even if only very roughly, the individual's various planes of construction.

No eggs of early reptiles have ever been found, and yet the most ancient reptilian remains (very fragmentary) date back approximately 338 million years, that is, to the Carboniferous period. (The *Hylonomus,* the most complete ancient reptile ever found, dates back approximately 300 million years.) Therefore, amniotic eggs must be at least that old. But it will take some time and study before paleontologists are able to fill in the details of the story.

The life of reptiles may have been difficult in the beginning: such a revolutionary transformation could only have taken place in small groups subjected to strong selection. But once it passed the test, this novel reproductive scheme spread rapidly because of the important advantage it offered: animals were no longer tied to water. It was possible to live (and reproduce) outside of swamps and lagoons: in forests, in the savannah, and even in the desert.

Thus, in this time of transition from the Carboniferous to the Permian, the egg is life's key to the environments and ecological niches of a planet that is becoming increasingly rich in resources and is encouraging new evolutionary experiments: reptiles.

THE EVENING OF DECEMBER 5TH (275 MILLION YEARS AGO)

Amphibians and reptiles: the same but different

Reaching the foot of a hill, we find a small bubbling stream shaded by the green fronds of a bower of tall trees. The sun can barely filter through. Small rocks create bottlenecks in the stream, forcing the water to flow into shallow transparent pools. Clumps of horsetail, which look like reeds with tufts growing at the junctures of its rings, sprout along the banks between the stones.

Taking off our shoes, we wade into the cool, refreshing water. What a pleasure to feel the chilly water swishing around our feet and ankles, taking the edge off the damp forest heat. The gurgling of the water accentuates the silence around us: not a sound can be heard, not even the call of an animal. And yet this place is full of life.

Only upon careful examination of the rocks on the banks, do we finally discover a round motionless body: a miniature alligator, no more than a foot and a half long. Its short, stocky body is plated with armor like a medieval knight. It has very sharp teeth and a nasty look. This animal is the terror of the area.

It definitely heard us coming long before it saw us: its hearing is particularly acute. In fact, it has fossettes behind its eyes with membranes stretched across them to form tympanic cavities that transmit sound vibrations. The animal looks like a reptile, but it is actually an amphibian. Its name is *Cacops.* It is, of course, a predator, like all amphibians.

The first reptile we see is lazing in the sun not far from the stream. This huge, peaceful salamander-like herbivore is called *Casea.* Its size has probably saved it from the *Cacops.* In fact, it has a small head and a large, broad body that looks as though it has spread out to be able to soak up as much heat as possible from the sun.

Unlike amphibians, reptiles no longer have to keep their skin moist; breathing is performed wholly by their lungs and there is no risk of dehydration. They even profit from lying in the sun: the external temperature raises the internal one.

Warm-blooded animals, so called because they maintain a constant body temperature, have not yet appeared on Earth. But the metabolism and therefore the physical efficiency of cold-blooded animals depends on the amount of heat they manage to incorporate. This is why lizards and snakes enjoy lying in the sun, in particular, on hot stones.

In the meantime, the *Cacops* has slipped into the water like a tiny crocodile and disappeared.

Independence from water

Superficially, then, amphibians and reptiles are almost undistinguishable in this period; they look like variants of the same great family. But if we observe them more closely, differences immediately become evident—differences that are quite remarkable.

Reptiles have introduced numerous biological innovations, some of which are fundamental. First of all, as we have already seen, the amniotic egg, which allows them to reproduce anywhere. Second, a skin that no longer has to be kept moist, a kind of sealed container, allowing them to live permanently away from water. These are actually the two innovations that made conquest of the continents possible: *Eusthenopteron* may have been a pioneer, with amphibians also breaking a lot of ground, but reptiles were the first animals to reach real independence from water. And they are producing still further evolutionary changes. Their skeleton is transforming, becoming more mobile and more adapted to movements on dry land. Probably the structure of the heart is also starting to change: the arterial blood (in the heart of modern reptiles) mixes only partially with venous blood thanks to a different partition of the ventricles (a process that will be completed in mammals).

The kidneys are coming into operation: ammonia is being transformed into urea and animals no longer have to take in large quantities of water to dilute it.

As in all primitive groups, the evolutionary shrub is extremely complicated and intertwined: the transition from amphibians to reptiles took place through a labyrinth of variations and adjustments, which were then put through the environment's selective filter.

Now that life has colonized the continents, climatic changes are taking on great importance. Variations in temperature and humidity; seasonal cycles; the influence of glaciers, rain, wind, and drought; climatic differences between the coast and the interior, between the highlands and the plains are a real change from life in the water where the environment is much more stable and "protected" from climatic fluctuations.

For a better understanding of the rest of the history of reptiles and amphibians and the role played in their evolution by climate (causing diversifications, traumas, new developments, and extinctions), account has to be taken of this parallel history, in particular, the oscillations in temperature and humidity that, like an unseen director, strongly conditioned evolutionary changes, and the climatic changes that led, in the end, to the appearance of dinosaurs and mammals.

The dimetrodons spread their "sails" and "fill up" with solar energy before nightfall.

11

Before Dinosaurs

DECEMBER 6TH (270 MILLION YEARS AGO)

In the highlands north of the equator

As a first step in understanding how climate has affected evolution, we have come to this plateau north of the equator.

The area is beautiful, rich in vegetation and water, much less hot and humid than the equatorial zone we have left behind. We can finally breathe again, no longer confined to that sweatbath known as the rain forest.

This is a region that knew no amphibian or reptilian life until not long ago. Indeed those animals were quite happy to stay in their sauna: the Carboniferous period had accustomed them to that kind of environment. Here, to the north (as to the south) of the equator, there were only insects, spiders, and millipedes. More importantly, no vertebrate had ever set foot on the plateau where we are standing right now.

But things have changed. The first animals to appear were small insect-eating reptiles, which found an unexpected banquet in a place almost void of predators. Given that they no longer depend on water for reproduction (they are now endowed with the amniotic egg) and breathing (a dry and resistant skin), these small reptiles were able to venture out of the equatorial sauna and develop adaptations for a drier climate.

Initially rivers and streams provided a semi-humid environment for their ascent to the highlands and plateaus. Even some amphibians chose this route. In fact, at the end of the last chapter we met two such explorers—the *Cacops* (an amphibian resembling a crocodile) and the *Casea* (a kind of giant salamander)—near a stream in a hilly area. But greater adaptations that only reptiles (and not amphibians) could develop were needed to abandon the wetlands forever.

The "sail" arrives

Paleontological excavations have revealed that animals with a huge bony crest on their back—a kind of enormous fan made of skin stretched over bony "ribs"—lived on the plateaus to the north of the equator at the beginning of the Permian period. Although some things still have to be cleared up about the exact functioning of this fan (or "sail," as some call it), most scientists believe that it was used to regulate temperature.

This was quite an improvement over the technique employed by the *Casea* (which had developed a large, flat back to expose as much of its body as possible to the sun). The sail was much more effective: a much greater area of skin with a dense network of blood vessels just under its surface was exposed to the sun. For a reptile in need of external heat (in order to improve its internal metabolism and therefore its movements), this sail was a cumbersome but valuable solar panel.

Fossils show that there was at least one animal in the Carboniferous that used this kind of structure successfully: the *Ianthosaurus*, an insect-eating reptile that grew to about two feet in length. Its sail was so successful that its descendants kept and enlarged it for millions of years.

Fossils also disclose that the *Ianthosaurus* was pursued and hunted (whether occasionally or regularly is unknown) by another reptile, the *Haptodus*, a three-foot-long carnivore. The fact that the *Haptodus* did not have a sail, but that its descendants later developed one, attests to the efficiency of this innovation.

Solar panels were not the only means of adapting to the environment, but they were certainly one of the most original and effective means of improving physical performance on dry land. Proof is that it was adopted by many dominant animals of that time, not only in the highlands, but also on the plains (even an amphibian, the *Platyhystrix*, developed its own personal version). But let's take a look at the kind of creatures populating these plateaus north of the equator during the Permian period.

The lake of the edaphosaurs

Leaving a forest of ferns, we climb a small knoll offering us a magnificent view of the surroundings. Below us is a large lake nestled into gentle, rolling hills.

The view is breathtakingly beautiful. The soft shores of the lake form tiny inlets and beaches covered in lush vegetation. The rays of the sinking sun bend in myriad directions on the waves' ripples.

Sitting down to admire the scenery, we raise our hands to shield our eyes from the sun. This should be an ideal place for highland reptiles. And, in fact, camouflaged in the grass are two edaphosaurs with their extraordinary sails.

The *Edaphosaurus* is a large lizard with a round, heavy body that can reach up to nine feet in length. It is a peaceful herbivore with a capacious digestive system. The sail is spotted with colors ranging from green to orange: these colors blend in well with the environment, making the animal less visible. In fact, the *Edaphosaurus* is a very vulnerable reptile. Despite its size, it has no effective defense mechanisms.

The two edaphosaurs lie motionless side by side, making use of every square inch of their sails to soak up as much sun as possible. They are "filling up" with energy before nightfall. Someone has calculated that reptiles of this size, weighing approximately 450 pounds, can raise their internal temperature from 79 to 89 degrees Fahrenheit by lying in the sun for an hour and a half (it would take them three and a half hours without the sail). The sail can also have the opposite function, that is, it can dissipate heat (as elephants do with their large, thin ears); if, in fact, the internal temperature of the animal is too high, all it has to do is to hold its sail perpendicular to the sun or expose it to the wind to eliminate excess heat. This kind of thermoregulation makes it possible to optimize heat as required, like an air conditioner.

This size of the sails suggests that the climate must have been mild at night, too. If it had been cooler, the sail would have brought the reptiles' blood temperature and, consequently, its body temperature down too quickly. But physiological remedies to such a problem could have existed: for example, contraction of the capillaries, decreasing blood flow to the sail, or nightly minilethargy.

As we contemplate these possibilities, we notice something moving on the opposite lakeshore, about two hundred yards away.

A mouth full of daggers

Through our binoculars, it looks as though a baby edaphosaur has been attacked by a much larger animal which has it by the nape of the neck. The young one is putting up a valiant but futile fight. The struggle goes on and the two disappear behind a bush. The tops of the ferns shake violently. Not far away an adult edaphosaur flees as fast as it can.

Now the predator reveals its identity: it is a *Dimetrodon*. Raising its head from the prey, it displays bloody jaws lined with strong, sharp daggers. A scrap of meat still dangles from the corner of its mouth. Its front paws rest on the body of the little edaphosaur, which now lies motionless, its neck lacerated.

The dimetrodon is a large reptile: about nine feet long. But it is also extremely aggressive, thanks to its voracious appetite, and the row of lethal daggers in its mouth.

Shaking its head back and forth, the dimetrodon greedily gulps down the bit of dangling meat without chewing it. This is typical of reptiles, as they have sharp cutting teeth but no molars for chewing.

The dimetrodon bends down over its victim once again, tearing another piece of meat away with such force that the little body is jolted from the ground. Now, as it swallows, it turns sideways, displaying its magnificent sail, which is shaped differently from that of the edaphosaur: it is higher—it must be approximately three feet at the center—and shaped like a wave.

Something now seems to be moving behind it: another dimetrodon hurrying to get in on the kill. But the new arrival does not receive a very warm welcome; the hunter is not about to share his booty. Yet the newcomer will not be driven away. Snapping at the rear part of the corpse, it dashes away with a large morsel hanging from its mouth.

Surveying the area through our binoculars, we notice two

more edaphosaurs grazing quietly in the vicinity. They do not seem to be frightened by or even interested in what is going on. Violent death is obviously rather commonplace in these parts; it strikes suddenly, like lightning, and is over just as quickly. But now the predator has had its meal and has sated its hunger, so at least for today no one else need fear attack.

Then again, no one seems to be very concerned about the drama that has just taken place. The sun will continue to heat the sails of the edaphosaurs and the waters of the lake will continue to reflect the sun's light for millions of years to come—until paleontologists come upon the fossil remains of the predator and its prey in a dried-up lake in Oklahoma.

The continents draw closer: the climate changes

In this early part of the Permian period, reptiles are starting to dominate the highlands and the areas to the north and south of the equator, where the drier climate favors their expansion. The amphibians, on the other hand, continue to live in their hot and humid equatorial steam bath.

But something is changing, something that will cause dramatic upheaval and mark the end of the amphibian era: a catastrophe similar to that which eliminated the dinosaurs, with the only difference being that some species of amphibians will survive this and subsequent crises and continue to live to the present time.

What was this dramatic event? To understand what happened, we will have to imagine the earth as seen from space. If an artificial satellite had photographed our planet at that time, it would have recorded an evolving geography. Laurasia and Gondwana (the two supercontinents comprising, more or less, the lands destined to become, respectively, North America, Europe, and Asia; and Africa, South America, Australia, and Antarctica) are approaching each other. The nearing of the two continents will end in "fusion," creating a supercontinent scientists have dubbed Pangea, which will in turn gradually split up at the time of the dinosaurs into the seven modern continents.

The most important change now occurring in the Permian, however, is that the lands in the southern hemisphere are gradually shifting northward toward the tropics and heating up.

In this period, our planet has two polar ice packs (which formed in the Carboniferous period): a smaller one in the north and a larger one in the south, created by the large glaciers that have formed on the continents. The northward shift of the southern landmasses is causing the glaciers to melt and triggering major climatic repercussions.

The equatorial belt is no longer that hot and humid paradise that it was for tens of millions of years (during the Carboniferous). The climate is starting to change, bringing increasing instability to the equatorial zone. Average temperatures will remain high, but wet periods will begin to alternate with dry ones.

These changes will gradually affect life, creating problems for the reptiles that have stayed on in the equatorial area and, above all, for the amphibians, which need water. This crisis peaked halfway through the Permian period, approximately 260 million years ago.

THE EARLY HOURS OF DECEMBER 7TH (260 MILLION YEARS AGO)

The destruction of a Garden of Eden

Twenty million years have passed since we explored the equatorial zone and witnessed the spread of amphibians in the early Permian. The swamps, lakes, rivers, lagoons, and bogs were then populated by perfectly adapted amphibians and reptiles.

Twenty million years later, the landscape has been transformed. It is the height of the dry season and drought has devastated the environment. Rain is no longer regular and the lack of precipitation has altered the vegetation and dried up many of the wetlands that dotted the region. This has had dramatic repercussions for amphibians.

Returning to the lake where we saw a *Diplocaulus* (the amphibian with a hat like Napoleon's) and an *Eryops* (the crocodile-like predator that attacked a small shark) at the beginning of the Permian, a sorrowful sight meets our eyes: the lake has almost completely dried up. Its bed is cracked and dry. The air is full of buzzing insects searching the mud for organic remains. Torrid heat rises from the bottom of this once beautiful lake surrounded by thick vegetation and tall trees.

But the most pitiful sight of all is at the lake's center: the small remaining pool of water is overcrowded with surviving amphibians—like shipwrecked passengers in a lifeboat. Some are half under the water, others are lying on its edges. Others are already dead and rotting in the sun. Skeletons lie everywhere.

These amphibians are different from the ones we saw during our first trip here. But all are condemned to death: the merciless sun in a cloudless sky forecasting no rain is evaporating the little water left, inch by inch. The lake is disappearing like water draining slowly out of a sink.

This is the way paleontologists found the fossilized skeletons of 260 million years: massed together in a very tight space. A catastrophe which in some ways recalls the crisis at the end of the Devonian period when freshwater fish were faced with the depletion of their food supply. But this one is much more serious. How are the amphibians (and the reptiles) reacting to this great crisis?

Adaptations, migrations, extinctions

There are generally three kinds of responses to profound environmental change: (1) adaptation, (2) migration, or (3) extinction. And there are obviously various combinations of these three responses: adaptation depends on the speed of the environmental change; migration may be a bridge toward new adaptations; enormous numbers of individuals may sometimes be wiped out, while the species itself survives.

All these changes are occurring in the second half of the Permian. Some amphibians are going extinct; others are surviving after many of their number have been decimated; some are adapting, others are migrating, looking for new niches.

One kind of adaptation developed by amphibians is particularly interesting: *encystment*, meaning that the animal creates a capsule around itself and waits for the rainy season. The same solution has been adopted by some fish (e.g., the modern protopteron) which, after the rivers and lakes dry up, encapsulate themselves and go into hibernation, as it were, until they are brought back into action by renewed precipitation.

One of the amphibians of the Permian that has developed this adaptation is the *Isorophus*. Fossils of about one hundred of them, rolled up and encapsulated, have been found in Oklahoma.

Things have been less dramatic for reptiles. Many have died, but their independence from water (made possible by scaly skin that does not need to be kept moist, and a means of reproduction—the amniotic egg—that can take place away from water) will allow many others to develop new adaptations in more arid climates. In fact, the heating of the land in the southern hemisphere caused by the shift toward the equator is opening up new terrain and creating new possibilities for migration.

New reptiles well adapted to the new climate are already beginning to appear: the therapsids. Oddly enough, their story is little known. It may be because dinosaurs seem to monopolize attention when the conversation turns to prehistoric reptiles. But therapsids represent an extraordinary moment in the evolution of life, not only because they were the rulers of the planet at one time and developed incredibly complex forms (probably already warm-blooded), but also and above all because they are the forefathers of mammals; that is, they represent the line which (after the demise of the dinosaurs) will give rise to modern mammals and finally human beings.

This is a tale really worth telling.

"The *Estemmenosuchus* lowers its head to drink. It is very unusual-looking, but the most striking thing of all is its head: horns sprout directly from the cheeks and nose like some kind of fantastic mask."

12

The Unknown Tale of the Therapsids

THE MORNING OF DECEMBER 7TH (258 MILLION YEARS AGO)

Waiting up in a tree

The tree we are sitting in overlooks a bend in the river. The current is a little less impetuous here, but what is more important, there are large, distinct footprints in the mud below us. The tracks are so numerous that it looks as though a herd of animals has passed through; bushes have also been uprooted and the vegetation trampled.

Looking at the tracks more closely, we notice that they do not all head in the same direction. This stretch of river must be a ford or a watering hole and a large herd of animals must have stopped here not long ago, perhaps only the day before.

That is precisely why we have climbed this tree: in the hope that the animals will return for more water. Animals that live in groups are generally herbivores and there is a lot of vegetation here. It seems reasonable to suppose that the abundance of food and the easy access to water might persuade the herd to remain in the vicinity for a while. And as this is watering time, all we can do is wait.

The sun has just come up and is filtering through the trees on the other side of the river. The morning is misty and chilly. Water has made the vegetation down by the river lush but the trees thin out just a little farther on. Beyond the woods, low ferns create a "meadow" effect: there is no grass in the Permian period and it will be a long time still before it appears.

We have been waiting for over an hour now, but have not given up hope. It shouldn't be long. If the herd follows the same path as yesterday, it should pass right below us, giving us front row seats.

Judging from the size and depth of the footprints, the animals must be very large. We can hardly wait to see them at close range. Now something seems to be moving on the other side of the river. We can't see anything, but we can hear some rustling in the reeds. It certainly doesn't look like a herd arriving. This is probably just a lone animal coming down to the watering hole.

After a brief pause, a head pops up above the vegetation: this

An *Eotitanosuchus,* a carnivorous mammalian reptile about seven and a half feet long.

is a large animal, at least seven feet long—an *Eotitanosuchus.* The appearance of this carnivorous therapsid is just what one would expect from a ferocious predator: it looks terribly nasty with its long, sharp fangs. The way it walks leaves no doubt about the fact that it considers itself the undisputed ruler of this territory.

But an *Eotitanosuchus* needs to drink as much as any other animal. And that is what it is about to do. Its front legs sink into the soft mud as it determinedly plunges its huge snout into the water.

The *Eotitanosuchus*, like all reptiles of its time, is incapable of sucking up water. It has to take the water into its mouth and then raise its head to let it run down into its stomach. It is a rather awkward performance that must be repeated again and again, but it serves the purpose.

Finally, the *Eotitanosuchus* has had its fill. Looking around, it shakes its head slightly and then slowly turns and disappears into the vegetation.

A herd at the watering hole

Only after another hour does the herd finally arrive. We are just on the verge of climbing down from the tree when a dull pounding on the ground and the rustling of ferns becomes audible.

All of a sudden, two large animals appear out of nowhere almost retracing the tracks left the day before. Others partially emerge from the vegetation behind them. There seem to be about ten in all, adults and young. These animals are estemmenosuchians.

Most unusual-looking, they are about as big as buffaloes, with shorter and sturdier legs and thick tails like reptiles. But the most striking thing of all are their heads: horns sprout directly from the cheeks and snout like some kind of fantastic mask.

The *Estemmenosuchus* (like the *Eotitanosuchus* which appeared earlier) is an archaic form of therapsid, but it already manifests some of the innovations that are evolving. Its skin, for example, is no longer scaly like that of a reptile, but smooth.

The fossils remains of these animals include not only bones and crania, but also bits of skin. Examination of the internal surface of this skin has revealed traces of numerous glands. Therefore, the skin must have been able to perspire and give off odors. And we now know how important odors are for mammals in sexual signaling, recognition, and marking territory, among other functions.

No odor wafts up to our vantage point, but it is nevertheless obvious that these animals are very different from all the reptiles we have seen up to now.

The herd cautiously approaches the river; several animals immediately lower their heads to drink, while others stand guard. Do they perceive the presence of a predator? The *Eotitanosuchus* has not appeared again, but it could be hidden somewhere awaiting its prey.

In any case, this herd of estemmenosuchians looks very powerful and aggressive: it would not be too easy for a solitary predator to attack such a big group. The habit of moving in herds is, in fact, the defense chosen by many therapsids. But certain predators have found a way around that as well by banding into groups and attacking simultaneously to disorient the prey and make them more vulnerable.

The estemmenosuchians continue to drink in turn. It is going to be a long day and the sun is already rather high in the sky. But these animals give us the distinct feeling that the direction of life is about to change. It is, in fact, in this shrub of forms that the transition toward mammals begins. That is why the therapsids have been defined as "mammalian reptiles": at a certain point in their evolution, they slowly began to develop characteristics that would be typical of mammals—constant body temperature (characteristic of so-called warm-blooded animals), skins and furs adapted to colder climates; maturation of the egg inside the body (culminating at delivery); parental care; and, finally, breast feeding of the young.

But the road will be long and complex and has not yet been completely retraced. The therapsids branched into a particularly complex genealogical tree. Originating from the sphenacodons (a branch of the numerous pelicosaurus family which includes the dominant reptiles of the Permian and all the animals encountered in the most recent part of our journey), the therapsids gave rise to various groups that we will eventually meet: in addition to the eotitanosuchians and the dinocephalians, to which the *Estemmenosuchus* belongs, there are the dicynodons, therocephalians, and cynodons (the group which is believed to have generated mammals).

The small herd grazes as it moves away, horns and other protuberances bobbing up and down as they disappear among the ferns and horsetails.

THE EVENING OF DECEMBER 7TH (253 MILLION YEARS AGO)

Tanks with crests and horns

Heading southward, we have crossed the equator and are now making for the land mass that is to become modern South America.

Actually, all continents were linked at the end of the Permian, forming the supercontinent known as Pangea. At that time, South Africa and South America were one and fossil remains of the same fauna on the two continents prove it.

The *Scutosaurus*, a large primitive reptile measuring up to seven and a half feet feet, would be the boast of any "monster designer."

As we move away from the equator, the climate becomes drier and cooler, especially at night. Analogous zones north of the equator have the same kind of climate, and it is likely that there have been exchanges or migrations of animals between north and south. Some seem to belong to the same stock. For example, these odd-looking "tanks" grazing in the clearing before us, pareiasaurs, closely resemble the scutosaurs found in the north.

These huge beasts look so menacing that the very idea of coming face to face with one of them is frightening. They are heavily armored with plates and studs, and their heads, in particular, have all kinds of bony crests, horns, and protuberances that any "monster designer" would be proud of. In reality, though, they are peaceful herbivores.

Yet these animals are not therapsids; they are among the few primitive reptiles, still cold-blooded, that have managed to adapt to these latitudes. They probably possess some internal characteristic that allows them to live so far from the equator. Whatever it is, their adaptation has been highly successful and will allow them to survive into the Upper Permian, for 10 million years.

An encounter with Moschops

Traveling a little further south, we finally reach the kingdom of the therapsids. These are young lands, only recently opened up to life: for over 100 million years (that is, from a time before the conquest of dry land by vertebrates) these lands had been covered with ice. They lay near the Antarctic Circle and gradually, as a consequence of the slow movements of the earth's crust, were pushed northward into warmer regions. The glaciers started to melt and this freed more and more land for colonization: first by plants, then by invertebrates, and now by mammalian reptiles—all forms that have been selected to live in climates quite different from the hot and humid equatorial zones.

In the course of less than 120 million years (a relatively short time, if compared to the preceding 3.5 billion), life has taken some very important steps in adapting to new environments. First, it left the seas, then it became independent of water and

The *Moschops* looks rather like a sumo wrestler. Well set on its sturdy legs, an adult can reach lengths of nine to fifteen feet.

later of wetlands. Now, with the therapsids, it has emancipated itself of hot climates as well.

Our first encounter with a therapsid in the southern hemisphere occurs on the edge of a small forest. The air is crisp and clear. It has just stopped raining and the ferns and other vegetation are still dripping wet. A few clouds are gathering in the distance behind an extinct volcano, but the whole scene is warmed by a gentle sun.

Grazing in a small valley not far from us is a small herd of *Moschops.* A large male stands apart to observe us warily. We remain motionless so as not to put him on edge. The huge beast looks like a sumo wrestler: well set on its sturdy legs and body is a swarthy neck supporting an enormous head.

The thickness of the frontal bone of the cranium suggests that this giant (which varies between nine and fifteen feet in length) fights its battles by using its head as a ramrod. This means that group life is regulated by struggles for dominance and the possession of territory and females, exactly as in modern times among mammals that live in groups.

Moschops are herbivores with hoe-like teeth well suited to procuring the enormous amounts of plant matter needed to keep up their muscle mass. They are peace-loving herbivores, but not animals to be provoked because when a head like that is driven forward with the force of such an enormous body, it can cause a lot of damage.

Predators attacking in groups (like the *Lycaenops,* "wolf snout," which inhabits these parts and has saber-sharp teeth) may be the only ones able to bring down a *Moschops,* striking all parts of its body—especially the more vulnerable ones—at the same time.

We decide to move on. Only when we have put some distance between ourselves and the herd does the dominant male turn around and slowly head back to the others.

DECEMBER 8TH (248 MILLION YEARS AGO)

A mini-hippopotamus

The aim of our trip to the south is not only to verify the variety of reptiles now inhabiting the diverse environments, but above all to reach the most recently populated lands, the places where the therapsids have developed adaptations to colder climates. And to observe such mammalian reptiles as the dicynodons or even the cynodons.

That is why we have set out for a hilly region slightly farther to the south, where the climate is considerably cooler and rainier than anything we have encountered so far. Survival here requires a very efficient metabolism as well as thermal regulation. Consequently, animals are likely to be warm-blooded.

This is the land of the dicynodons, reptiles very similar to mammals. In this region in particular, the *Lystrosaurus* rules.

Following a stream that zigzags through the hills, we finally come upon a few lystrosaurs close to a small lake. Some are

The *Lystrosaurus* looked like a small hippopotamus (three feet in length) and, like modern hippopotami, enjoyed the water.

grazing on its banks, others are standing half-submerged in its waters. Short and stout, no more than three feet in length, these animals look like small hippopotami.

It is obvious from their build that they are well equipped for the cold. Their bodies are round—a shape that allows minimum heat dispersion—and their tails are short and compact.

The *Lystrosaurus* is a herbivore and has a curious beak-like snout with which it severs plants and shrubs. It also has two canine teeth (hence the name "dicynodon," literally "two-dog-tooth") which it probably uses for defense.

Another lystrosaur climbs agilely out of the water in spite of its stocky build. Like hippopotami, these animals love to be in the water and the shape of their head confirms this: both their eyes and their nostrils are located on the upper part of their head and nose. This allows them to see and breathe while completely submerged, like hippopotami.

Lystrosaurs are almost certainly warm-blooded: all the features of their body and the environment in which they live suggest it. The key to living successfully in places like this is having an internal "heating" system, which makes them independent of outside temperatures. But warm-blooded animals also have more active metabolisms, which means they require more food (and an adequate body structure).

Watching these hippopotami-reptiles paradoxically brings to mind a very ancient image: the protocells that we met in the primordial seas at the beginning of our journey. They basically had the same problem: they only managed to "live" thanks to the chemical reactions that took place inside them and provided them with a primitive "metabolism." And they only survived through time thanks to their ability to duplicate themselves by means of a structure known as DNA.

That is exactly what is happening among these lystrosaurs right now: while most are grazing on the shores of the lake to fuel their heating systems, one male has given chase to a female and has now climbed onto her back and is inseminating her. In the thousands of variations introduced by evolution, metabolism and reproduction are still the two irreplaceable pillars of life. It is the same in the Permian period as in the era of technology.

Wolf-reptiles

We have come a long way to meet the last (and perhaps most mammalian) therapsids, but it has been worth the effort. Animals like the *Cynognathus,* a large carnivore that hunts in these temperate zones of southern Africa, almost look like modern animals from afar.

Through our binoculars, the *Cynognathus* on the other side of the valley resembles an "assembly" of various animals. The body already has a mammalian appearance (a short chest), while the tail is still reptilian. The legs and, above all, the head resemble those of a dog or a hyena. It has thick, muscular lips and modern mammalian-type teeth: besides the canine fangs, it has incisors and sharp multicuspid molars like a dog.

The *Cynognathus* probably occupies the niche of the wolf: it hunts in a group and behaves gregariously. Two more *Cynognathus* are, in fact, loping over the top of the hill to join the first. Their gait is no longer reptilian; their hind legs are straight and upright under their body, giving them a completely different stride. Their bodies seem to be covered with fur.

What do these primitive wolves hunt? Fossil remains indicate that the *Kannemeyeria* (another therapsid) was probably among their prey. A nine-foot-long herbivore, this short-legged bison with a large beak was a larger version of the *Lystrosaurus,* the small reptile-hippopotamus that we saw during the last part of our journey.

The *Cynognathus* is of course not the only predator in this part of the world. Many other forms of therapsids have specialized in hunting. The variety of shapes and sizes demonstrates the abundance and diversity of the therapsid evolutionary line. Some therapsids (the therocephalians) have developed special characteristics that will later become part of the genetic heritage of mammals: the nasal concha, that is, the labyrinths in the nose that increase the surface area for olfaction as well as for the exchange of heat, humidity, and air, and the palate, which makes

A pack of *Cynognathus* attacks a *Kannemeyeria,* an immense herbivore measuring over nine feet. *Cynognathus* probably occupied the niche of the modern wolf; it hunted in packs and behaved gregariously.

it possible to eat and breathe at the same time, something reptiles cannot do.

Another therapsid, the *Thrinaxodon,* also lives in this zone. It strongly resembles a dog, but has clawed paws like a monitor, and is an excellent and fast runner. It has a bone in its foot which will eventually become very important for evolution and give rise to the heel. In addition, it probably already has a diaphragm, that membrane in the chest which allows the lungs to pump much more efficiently.

This period is bursting with innovations and adjustments that are transforming these former reptiles (they may now be defined as such) into protomammals. But, as has happened so often in the story of life, things do not always work out as they should and certain prospects can change overnight. The forerunners suddenly find themselves in difficulty and "outsiders" overtake them.

This is what will happen to the therapsids: their seemingly relentless development will suddenly be brought to a complete stop by a merciless killer.

An ecological disaster

The killer is the climate. Once again, it is the climate that conditions and selects the life that is to evolve on Earth.

The great thaw in the southern part of the planet which had opened new areas and new niches for the therapsids has boomeranged. With the melting of the last glaciers (which have almost totally disappeared) temperatures have increased, throwing the therapsids, which are specialized for colder climates, into crisis.

These animals, which owe their success to the climatic changes that followed the Carboniferous period, are now forced to face environmental upheaval and a catastrophic crisis which will cause mass extinctions. The advantages of a mammalian metabolism have become a handicap: the therapsids had adapted to a climate that is now vanishing. Other reptiles favored by the climatic changes are beginning to compete with them.

Entire groups have disappeared, others are being decimated. Almost all the forms that we have seen in the last part of our journey have vanished.

The vegetation has changed, too. Plants with a seasonal loss of foliage (an important evolutionary gain) have disappeared, while those adapted to warm, dry climates, like palms, are making their appearance once again.

The extinctions are generally gradual, but in some cases they probably also take place in waves. Marine life has not been spared either: the general heating is causing a rise in the temperature of the water, affecting the microfauna. In a period of a few million years, 50 percent (perhaps as much as 75 to 90 percent) of marine life will disappear. This is an ecological catastrophe.

The situation is no better in the north. Ice is starting to encroach upon the land, creating devastating climatic fluctuations. In the end, warmth will win out in the north, too, and the ice will melt (there are no ice packs during the era of the dinosaurs, which is about to begin). But in the meantime, these changes will have terminated an evolutionary trend that had enormous possibilities. You might say they have shackled an athlete who was about to jump a high hurdle. The surviving therapsids (and their descendants) will be shackled for a long time. They will be no more than extras in a scene dominated by others—dinosaurs—for almost 170 million years.

But in spite of everything, these protomammals will manage to evolve into mammals. They will, however, live almost clandestine lives, marginalized, nocturnal, and limited to small individuals. Only after the dinosaurs are in turn thrown into crisis by drastic climatic change will these small mammals be able to regain the territories lost by their ancestors and flourish once again. And the resulting diversity of forms will know no limits.

Thus, the tale of the therapsids is drawing to an end, while that of the dinosaurs is about to begin.

DECEMBER 9TH (240 MILLION YEARS AGO)

The ancestors of the dinosaurs

The dry warmth of the Triassic period has spread over a vast part of Pangea, causing enormous tumult. As we have seen, some lines that seemed successful are growing extinct, while other that were formerly of minor importance are now coming into their own. Others are giving rise to totally new lines.

The period that extends from the decline of the therapsids to the rise of the dinosaurs is characterized by the intertwining of various evolutionary trends. The genealogical tree of life shows that evolution started off in various directions—like a diaspora—during this period.

Certain groups of reptiles gave rise to turtles. Others (such as lepidosaurians) generated lizards and snakes. But the most successful line seems to be that of the archeosaurs. The first archeosaurs, called thecodons, generated various groups which in turn gave rise to crocodiles, pterosaurs, and dinosaurs.

In this period, evolution branched out like a candelabrum, generating lines of reptiles that will in some cases descend to the present day (snakes, lizards, tortoises and turtles, and crocodiles). It also gave rise to those giant reptiles that will soon occupy all terrestrial niches for a long time to come. These creatures, which survived three geological periods (Triassic, Jurassic,

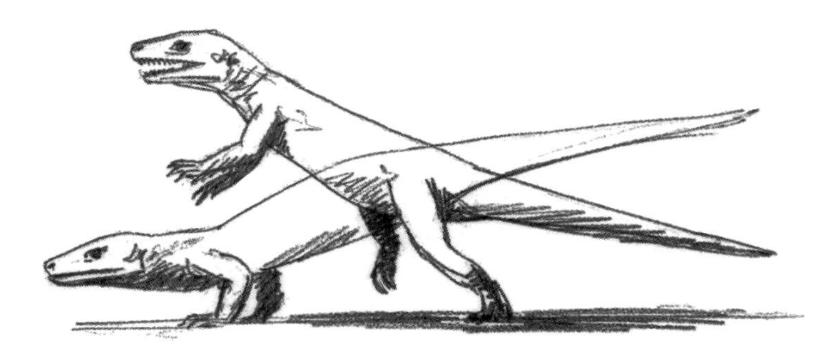

The *Euparkeria* is a reptile only two feet long, but it has sharp, saw-like teeth just like future carnivorous dinosaurs.

The *Lagosuchus* is tiny, no more than a foot long, and yet the pelvic bones of this long-legged lizard seem to indicate that it may well be the progenitor of all dinosaurs.

and Cretaceous), never cease to amaze and their story has filled entire encyclopedias.

How did dinosaurs evolve? Only a few facts are known about the origins of dinosaurs and their ancestors. Two ancestors were the *Euparkeria* and the *Lagosuchus*. Curiously, both are very small. And what is it that distinguishes them?

To a greater degree than any other reptile ever found, they have characteristics that are typical and unique to dinosaurs. The *Euparkeria*, for example, is only two feet long, but has sharp, flat, saw-like teeth just like those of future carnivorous dinosaurs. Its hind legs are well developed and upright under its body (a characteristic of all dinosaurs) and it has a posture which is partially if not wholly bipedal, typical of an entire line of dinosaurs.

The second probable ancestor—the *Lagosuchus*—is in front of us right now, walking spryly along the bank of a small stream. It is tiny, no more than a foot long; if it weren't for its long slender tail, it would fit into the palm of a hand.

It seems incredible that this minute creature could have given rise to all the great reptiles that ruled our planet, from the immense brontosaurus to the terrible tyrannosaurus and the colossal ultrasaurus. And yet the pelvic bones of this small long-legged lizard seem to indicate that it may well be the progenitor of all dinosaurs. Not only that, but it may also be the forebear of pterosaurs, those flying reptiles that filled the prehistoric skies. In fact, its pelvis has characteristics typical of both of the two great evolutionary lines of dinosaurs: the saurischians (which had reptile-like pelvises) and the ornithischians (which had bird-like pelvises).

We watch as the *Lagosuchus* walks away. It has a long way to go and the story which it initiates is absolutely extraordinary. So we'll try to tell it with a short excursion to the planet of the dinosaurs.

"What we have before us is one of the first real dinosaurs: the *Coelophysis*. . . . Small icarosaurs sail from one cycad to another. . . ."

13

The Planet of the Dinosaurs

DECEMBER 10TH (225 MILLION YEARS AGO)

The Triassic desert

The desert stretches ahead, hot, arid, hostile, and lifeless. After walking for hours we have behind us only barren rock and sun-scorched stone, and before us vast expanses of sand and dunes. Everything is motionless and silent. A hot wind burns our skin and bakes our surroundings.

Rain is obviously not too frequent here—or anywhere else for that matter. The global heating that took place during the Triassic period has exacerbated the continental climate of the inland areas of Pangea. In the absence of moisture-bearing winds and the influence of the sea, many regions have turned into deserts, with endless dunes and rocky plains.

The heating has not spared any geographic region: the temperatures at the poles, for example, are now between 50 and 60 degrees Fahrenheit.

The only areas covered by vegetation are, of course, those that enjoy precipitations and humidity. The coastlines and the areas skirting lakes, rivers, and swamps are covered with an extremely rich mantle of plant growth. Sometimes they look like oases in the middle of this desolate landscape. We are heading for one right now.

A green valley

Seen from above, the oasis looks like a broad green serpent winding its way to the horizon, or a kind of woody tunnel. Following the meandering river, luxuriant vegetation has created a green valley like that of the Nile in the Egyptian desert.

Descending into the valley, we pass lone trees and shrubs that have fronds like those of a cedar or ginkgo. Other spire-shaped trees have flat, ribbon-like leaves. The ferns and cycads, with their characteristic pineapple shape, plumed on top with fringed branches like palm trees, become increasingly dense as we approach the river. The buzzing of large beetles (the first sounds we finally hear) drowns out the drone of the other insects flying busily from plant to plant.

All of a sudden, a slender shadow takes off from a high branch and glides downward. With a flick of its long tail, it veers around 180 degrees, and, snapping up a beetle, continues on a new trajectory to another branch, lower down.

In profile, it resembles a lizard with wings. And drawing nearer, we see that it looks like a lizard at close range, too, only that some of its ribs have protruded from the sides of its body and have been covered with a membrane allowing it to glide through the air. This is an *Icarosaurus,* one of the first attempts at flight by the vertebrates. But like Icarus, this reptile will fail to conquer the air and will leave no descendants.

The *Icarosaurus* may not be the oldest gliding reptile: another very similar "winged lizard" found in Madagascar was probably able to glide just before the *Icarosaurus* appeared, but the dates still require confirmation.

It is important to underline that these gliding reptiles have nothing to do with the famous pterosaurs (which are actually able to fly by beating their wings). They belong to a completely different, though parallel, line.

Coelophysis *on the march*

Something seems to be moving behind a bunch of cycads. Very timidly and cautiously, the long muzzle of a reptile pokes out of the vegetation. Two large yellow eyes jerkily scan the surroundings. The head cranes forward, revealing a long, wrinkly neck. A row of small, flat, curved teeth, as sharp as razors, peeps out from below the upper lip: this is definitely a predator.

Its expression is fixed as if it were wearing a mask; only its eyes dart from side to side. The line of its mouth draws a sinister "smile." Again the head cranes forward and with a leap the animal pops out of the bush.

One feature immediately strikes us: this creature is a biped. It is the first permanent biped we have met. Its tail is held straight and high as a counterweight. Its front limbs are already much smaller. What we have before us is one of the first real dinosaurs: a *Coelophysis.*

More precisely, it belongs to the group of coelurosaurs which includes slender, exclusively carnivorous dinosaurs with small heads, long necks, and prehensile "hands." Another common characteristic is their speed and acceleration.

About the size of a human being, this animal almost looks like a bipedal monitor. Another head pops up from the vegetation, then another and another: this is an entire group of *Coelophysis*.

Now in full sight, the animals look around with short rapid head movements. Attracted by something in a shrub, one edges away from the rest of the group. The others follow it. With lightning speed, the first extends its neck and then leaps forward to capture a tiny reptile, swallowing it whole before its companions can tear it away.

These early dinosaurs are still very different from the giants that will develop later and become the absolute rulers of the continents. For now, dinosaurs are still rather rare and certainly not dominant. It will take millions of years before that comes to pass. To see the enormous creatures, we will have to move on to their moment of splendor, the Jurassic period. But first let's try to understand what makes these animals so special and unique. A family portrait should highlight their characteristics.

What is a dinosaur?

To gain an understanding of what a dinosaur is, it may be important to point out first what it is *not*. Dinosaurs are not lizards or giant monitors; nor are they snakes, turtles, or crocodiles. Much less are they birds or mammals. Yet, they have many features in common with these "relatives." They have scaly skin like lizards; they lay eggs like birds; they build nests like crocodiles; they take care of their young like birds and mammals. Many have beaks like turtles, others are probably warm-blooded like birds and mammals.

They are basically another variation on the vertebrates that exist in the world today.

So, how can a dinosaur be recognized? How can a paleontologist identify a bone as belonging to a dinosaur?

Well, dinosaurs have characteristics that distinguish them from all other reptiles:

1. They lived at a very precise time in prehistory: between 220 and 65 million years ago. That is, in the Mesozoic, an era comprising three different periods: the Triassic (250–205 million years ago), the Jurassic (205–135 million years ago), and the Cretaceous (135–65 million years ago). Any fossil found in older or more recent sediments cannot belong to dinosaurs.

This explains why no dinosaur bones have ever been found in Italy: at the time when the great reptiles dominated the earth, only a few points of the Italian peninsula had come to the surface in the form of isolated islands. Although a few tracks and bones have been found (including the skeleton of a small therapod fifteen inches long), they will always be exceptions; that is, they will belong to dinosaurs that lived on the edge of the great sea that covered Italy at that time or on the islands just off the coast.

2. Dinosaurs are land animals. That means that they do not fly or live in the sea, although they do venture into the water (like dogs), as fossil tracks in lakes and rivers demonstrate. The flying reptiles (pterosaurs) and long-necked marine reptiles (plesiosaurs) often found in books on prehistoric times are not dinosaurs, but merely cousins, like crocodiles.

3. Dinosaurs' legs are placed vertically under their bodies, like those of an elephant. This is the feature that really distinguishes dinosaurs from all other reptiles. Both bipedal and quadrupedal dinosaurs have this feature which keeps their belly—and their tail—from touching the ground.

Crocodiles, monitors, lizards, and all other reptiles have their legs set at an angle: "elbows" and "knees" are bent and point outward. One consequence of this is that their bellies touch the ground when the animal is at rest. This position looks somewhat like a soldier creeping forward in "prone" position. This is not a characteristic of dinosaurs. To imagine the way a dinosaur walks, think of an elephant, a rhinoceros, or a cow, not a crocodile.

A host of different species

Over 440 species of dinosaurs have already been discovered and new ones are being unearthed at a rate of approximately one every two months. Dinosaurs were recently found for the first time on the Arabian peninsula and in the Antarctic. This clearly indicates that they lived on all continents. And there were dinosaurs of all kinds: bipeds, quadrupeds, carnivores, herbivores, dinosaurs with teeth, dinosaurs with a beak, dinosaurs with teeth and a beak, and so on. Some were as long as three buses in a row, others were about the size of a cat.

How to classify them? All 440 species fall into two main categories (orders): the *saurischians* and the *ornithischians,* distinguished by the placement of the three pelvic bones (the iliac, the pubic bone, and the ischium).

Seen in profile, the pelvis of the saurischians resembles a hand with the thumb, the index, and the middle finger extended and separated. When seen in profile, the pelvis of the ornithischians, on the other hand, looks like a hand miming a gun (a double-barreled gun), with the pubic bone and the ischium united.

Of course, this very rudimentary method does not take account of some important differences in the shape of the bones making up the pelvis. But when we stand in front of a skeleton in a museum, it is a fast and effective way of recognizing a dinosaur.

Saurischians

The saurischians include all carnivorous dinosaurs (called theropods) such as the famous *Tyrannosaurus,* the *Allosaurus,* and the *Deinonychus,* a small and ferocious killer. Other saurischians (called sauropods), which include the largest dinosaurs ever to walk the earth, such as the *Brontosaurus* and the *Diplodocus,* were herbivores.

Ornithischians

The order of the ornithischians contains only herbivorous dinosaurs. In addition to having several sets of teeth, all probably had a beak at the end of their nose (or at least on their lower jaw). Ornithischians often had bony crests running down the length of their backs, which are sometimes still visible in the skeletons on view in museums.

The ornithischians are made up of five very different groups:

1. the famous stegosaurs, with the typical bony plates on their backs
2. the ceratopsids, living tanks with a huge bony shield at the nape of their necks and horns on their heads (Triceratops, *Styracosaurus,* etc.)
3. the ornithopods, meek herbivores of various sizes, mostly bipeds (hadrosaurs)
4. the ankylosaurs, heavily armored quadruped dinosaurs shaped like "overturned boats" with defensive spikes all over. Sometimes they had bony clubs at the end of their tails to defend themselves from theropods

Immense sauropods like the *Diplodocus* had tiny heads perched at the end of long necks.

5. the pachycephalosaurs, bipedal herbivores with very strange crania; in some forms, the cranial vault was extremely thick and distended out of all proportion, while in others it was reinforced by a crown of bony spikes. These adaptations probably served for charging other males during duels, like rams or the *Moschops* we met in the last chapter (the bones of these dinosaurs are so rare that it is believed that they lived in areas such as mountains unfavorable to fossilization).

The world of dinosaurs is so rich and varied that it is impossible to explore it in only a few pages. Three chapters will nevertheless be dedicated to these great reptiles to illustrate some of the key moments in their evolution and the evolution of the earth in this extraordinary era.

DECEMBER 16TH (160 MILLION YEARS AGO)

An encounter with a brachiosaur

We are in the middle of the Jurassic, the period of the giants. Sixty million years have passed since we met the *Coelophysis.* Pangea (the supercontinent) still exists, but it has started to "tatter" at the edges, with the seas moving onto the land. The climate has changed profoundly once again with the dry heat of the Triassic giving way to a hot and humid climate. Nevertheless, some immense deserts with enormous dunes remain.

The great forests have grown back, with strange-looking plants, some of which vaguely resemble our conifers. From the top of one of these, we have an excellent view of the forest around us, a kind of tropical rain forest that extends as far as the eye can see.

A storm is lashing out on the horizon. Lightning bolts zigzag down from black clouds discharging torrential rain. In the garden of Eden below us, reptiles of all shapes and sizes now graze and hunt: in the last 60 million years, dinosaurs have gradually multiplied and diversified to occupy all ecological niches.

Our lookout gives us a view of one of the great protagonists of this period: a gigantic brachiosaur grazing in the clearing below. This sauropod looks like the reptilian version of a giraffe: it stands thirty-six feet high and can weigh up to fifty tons. Like a giraffe, its front legs are much longer than its hind legs (the shoulder is about eighteen feet high). Its long limbs and long neck allow it to reach the upper branches and leaves that are inaccessible to other herbivores.

Leading the way with its head, the dinosaur sidles over toward our tree as if in slow motion. It looks like a living version of a fireman's ladder. Not noticing our presence, it nudges in

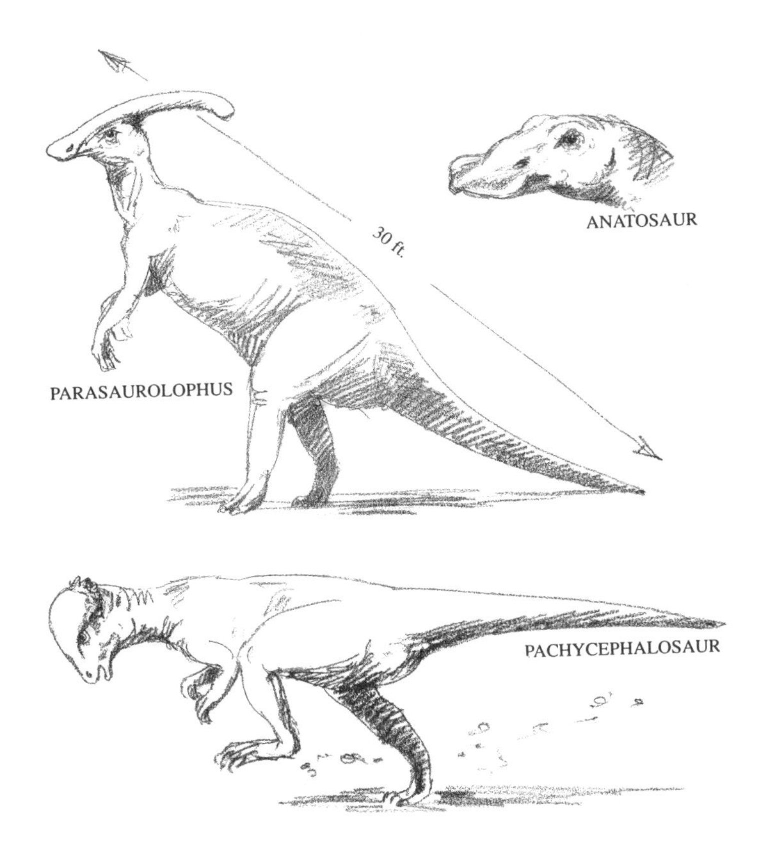

among the leaves to nibble not far from us, about forty feet above the ground.

Its small head is shaped like a hiking boot. It eyes are tiny and its nostrils are high-set and almost stuck between them. The broad mouth contains a row of cylindrical teeth that allow it to "rake" the fronds and tear away the succulent leaves.

We can hear its deep, strong breathing and feel its warm breath, heated by the fermentation of the plant matter in its enormous stomach. This internal "boiler" probably generates the heat needed for its metabolism.

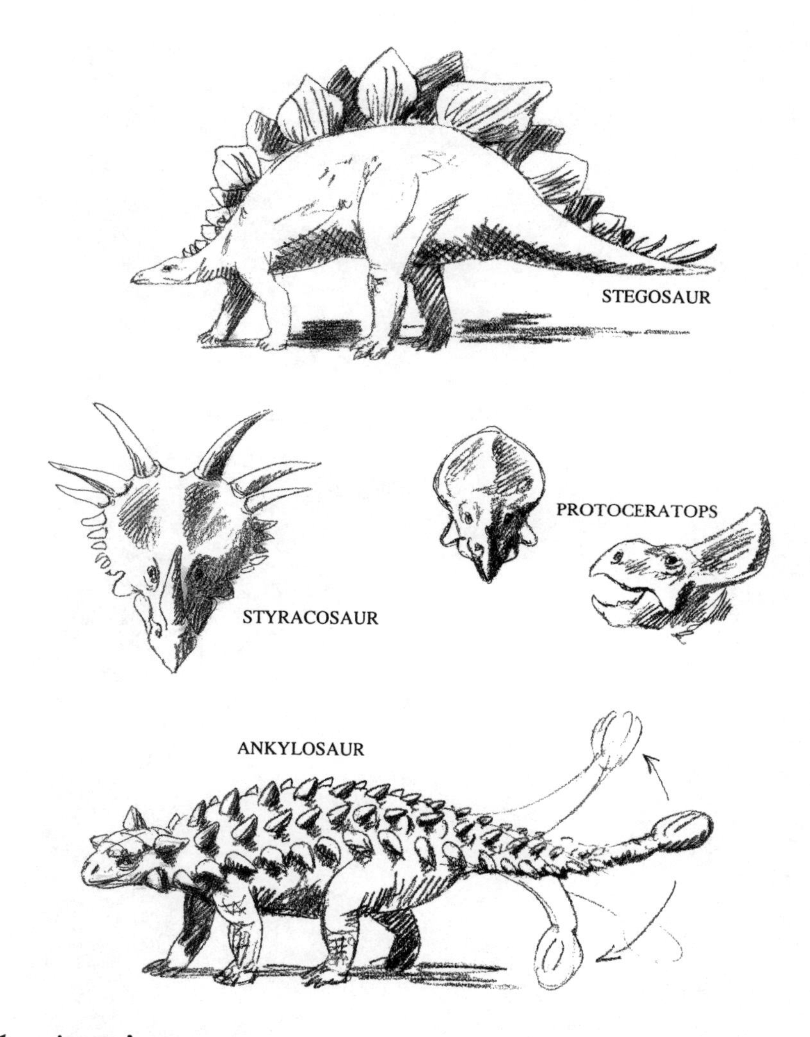

The giants' secrets

An artery is pulsing on one side of the huge animal's neck. The brachiosaur must have an extremely large and strong heart to be able to pump blood all the way up to its head forty feet away. And its circulatory system must have developed special adapta-

"Not noticing our presence, the brachiosaur nudges in among the leaves to nibble not far from us about forty feet above the ground."

tions, too. It may, like the giraffe, have blood that is richer in red blood corpuscles, a faster heartbeat, anti-return valves in the veins, and stronger arteries.

The brachiosaur has spotted us at last. Violently jerking its head back, it ambles away "in slow motion."

Other brachiosaurs are grazing around us, their bodies almost camouflaged by the tree trunks. One of them is even larger than the rest: it must be at least forty-five feet tall (i.e., as high as a five-storey building), 120 feet long (the length of three buses in a row) and as heavy as twelve elephants. Brachiosaurs are probably the largest and heaviest animals of all time.

The brachiosaur is only one representative of a vast group—the sauropods—which includes other long-necked dinosaurs. Sauropods include the famous *Brontosaurus* (or more correctly *Apatosaurus*), which is about sixty-five feet long, and the *Diplodocus,* which is ninety feet long. Studying the bone structures of their necks (and tails, which are supported by dozens of vertebrae), some experts have noticed that the distribution of the lines of force along their body is surprisingly similar to that of suspension bridges like the Golden Gate or the Brooklyn Bridge.

Watching these animals extend their long necks to the upper reaches brings to mind elephants as they extend their trunks. In some way, the neck of the brontosaur is the equivalent of the elephant's trunk: it is like a trunk with a head on the end of it.

Thanks to its special climatic and environmental conditions, the Jurassic period was the golden age of the sauropods; many (but not all) of them went into extinction with the advent of the Cretaceous.

THE EVENING OF DECEMBER 17TH (150 MILLION YEARS AGO)

Tearing up a stegosaur

In the last 10 million years, many things have happened in the forests and on the plains of this planet, which now abounds with life and is populated by large reptiles. While trekking into this place, we made a very exciting discovery: we found a fossilized skeleton of a brachiosaur in the midst of a fossil-rich area.

"The allosaurs sink their long teeth into the foul flesh of the half-submerged stegosaur."

The evolutionary time frame is so long that events occurring in the same geological era can be very far removed from one another. The Jurassic period alone extends over almost 70 million years (longer than the time that separates the modern era from the last of the dinosaurs). As a result, there was a lot of prehistory in prehistoric times, too.

Our hiding place behind the twisted roots of a fallen tree gives us an excellent view of a bend in a river. The swollen and putrid carcass of a huge stegosaur has been washed up onto the opposite riverbank.

The stegosaur, famous for the plates of armor on its back (probably used as thermoregulatory panels to modulate its internal temperature), is not the only cadaver in the area. Long racks of white vertebrae emerge from the waters or are half sunken in the sand; pelvises and femurs already covered with moss, gnawed or trampled by carnivores, protrude from the vegetation. These are all signs that this is a place where dinosaur remains washed down by the river accumulate. It is very likely that the bones of many animals buried at the bottom of the river have already been "packaged" for twentieth-century paleontologists who will find them perfectly fossilized.

Three broken teeth are lying on the ground not far away. They are flat and about the size of cigarette lighters. Their saw-like edges leave no doubt: these teeth belong to a carnivorous dinosaur. Indeed dinosaurs often lose their teeth (which are soon replaced) while devouring their prey, thus leaving a kind of "visiting card."

Before long, one of these predators—a ceratosaur—appears beside the dead stegosaur. About eighteen feet long, it is built like a classic bipedal carnivore: strong, nervous hind legs, and small, prehensile front limbs; sharp claws, a raised tail, and a large head, armed with two rows of well-honed teeth.

This solitary predator has one curious feature: a horn-covered bone in the shape of a blade on its nose. The animal does not use this flat horn to tear apart its prey, but, like other bipedal carnivores, to intimidate adversaries during mating battles. It could also serve for sexual recognition among members of the same group.

The ceratosaur starts to tear away and gulp down bits of decaying meat and skin from the stegosaur's carcass. Although a predator, like a lion on the savannah, it never turns down a free dinner offered by chance. But suddenly it raises its head and looks around, its red tongue palpitating in its half-open mouth. Whirling its body around to one side of the river, it threateningly bares its teeth.

Five allosaurs have emerged out of the bushes and are approaching along the riverbank. They are very similar to ceratosaurs but much larger (30 to 36 feet long). Instead of a horn on their forehead, they have two small crests over their eyes. Their three-fingered hands have long horrible claws.

Although no match for the newcomers, the ceratosaur makes a stand to defend its booty. But the allosaurs suddenly all plunge at the ceratosaur and the carcass. Just as the jaws of one of them is about to close in on the ceratosaur's neck, it wheels around and disappears into the dense vegetation. Who knows whether it will wait there in the hope of finding some leftovers after the allosaurs have finished banqueting.

Spraying water on all sides, the allosaurs sink their long teeth into the foul flesh of the half-submerged stegosaur. It is quite terrifying to witness the violence with which they tear at its meat. The mutilated bones are soon picked clean; not a morsel remains.

It is dusk when the allosaurs finally move on and the moon is already rising over the horizon. This moon is somewhat different from the one that shines down on us today: not only is it slightly larger (that is, closer), but its surface has not yet been modeled by the impact of meteors and asteroids. Above all, the most visible crater, caused by Tycho, is still absent.

Strange calls echo through the forest and across the river. A world similar to ours is slowly dawning.

"The *Archaeopteryx* looks like an evolutionary photomontage. Is this animal a reptile, a dinosaur, or a bird?"

14

The Advent of Flight

DECEMBER 18TH (145 MILLION YEARS AGO)

Feathers and quills

Large tree ferns similar to palms sway gently in the breeze blowing in from the sea. In this region, a dense network of channels and streams creates an archipelago of verdant islands and islets which gradually give way to large lagoons closer to the sea.

The shores are littered with the spiral shells of ammonites, descendants of the cephalopods that we saw in the Cambrian seas millions of years ago. Resembling hunting horns inhabited by octopi, forms of ammonites have developed in all shapes and sizes and have spread throughout the seas, especially along the coastline.

On dry land, on the other hand, a very special dinosaur—the *Compsognathus*—lives among the low ferns. The size of a chicken, this creature looks like a miniature version of the enormous bipedal dinosaurs. Fast, timid, it has adapted itself as a predator and is the "terror" of the insects and small vertebrates that share its habitat.

There is one right now, devouring a small reptile that it has just flushed out of the vegetation: a *Bavarisaurus,* that is, a gecko with a long tail that lives in these parts. Before it has even finished swallowing its catch, it sets off with a spry gait, the tail of its prey still wiggling in its mouth.

Something seems to be floating in the small channel beside us. It is not an ammonite shell, but something much lighter and more colorful. It's a feather! A feather very similar to those of modern birds.

It is asymmetrical, that is, the surfaces on either side of its central axis (rachis) are not identical: one is broader than the other—a characteristic that gives the feather a very efficient aerodynamic structure.

Any animal with a feather like this has to be a pretty good flier, not just a "gliding lizard" like the *Icarosaurus* which we saw in the Triassic period. This must be a primitive bird.

The Archaeopteryx

Our patient search of all the islets finally pays off. We finally come upon an animal about the size of a pigeon and unlike anything we have seen up to now: an *Archaeopteryx.*

The only suitable way to describe this animal is to think of a lizard with feathers. Or better yet, a *Compsognathus* with feathers. The *Archaeopteryx* has the head of a dinosaur, with tiny conical teeth fit for catching insects (no living bird has teeth), "hands" with strong claws, and a long tail like the *Compsognathus.* It looks like an evolutionary photomontage.

Is this animal a reptile, a dinosaur, or a bird? Interpretations have varied since the first fossils were discovered in 1861. Only seven findings, including skeletons and other remains and a lone feather, have been found of this reptile. Some scientists believe it to be the ancestor of birds, others an evolutionary dead end (in their opinion, birds descended from a primitive crocodile that lived in the Triassic or from some archeosaur that has not yet been found). The first *Archaeopteryx* was discovered only two years after Charles Darwin published his *Origin of Species* and was long used to confirm evolutionary ideas.

Debate about the evolutionary placement of the *Archaeopteryx* among birds is still very heated. About twenty years ago, an American researcher, John Ostron, came to an important conclusion after careful study of the fossil remains of the *Archaeopteryx.* He proposed that the origin of birds should be sought in a particular group of small bipedal carnivorous dinosaurs, the coelurosaurians, which comprise animals such as the *Compsognathus* and the *Coelophysis,* to which the *Archaeopteryx* bears a striking resemblance. The skeleton of the *Archaeopteryx* and those of these dinosaurs have about twenty features in common, including the bones of the cranium, the forelimbs, the shoulders, the back, the legs, the hips, and the tail.

Today, many researchers agree with this hypothesis or a variation of it which claims that birds do not descend *directly* from these dinosaurs, but share a common ancestor with them that probably lived about 200 million years ago.

In the end, the way things stand is that some scientists feel that the *Archaeopteryx* is a "flying dinosaur" (and that, in some ways, all modern birds should be considered as such), while others feel that it is only a "close relative" of the dinosaur.

The forefathers of flight

Does that mean that the *Archaeopteryx* is the first real "bird"? Well, it's not quite that simple. Some hypothesize the existence of an intermediate form, the *Protoavis.*

The discovery of some very ambiguous fossils a few years ago rekindled debate. Are they the remains of a *Protoavis,* that is, the ancestor of the *Archaeopteryx* and of birds? Or merely those of some primitive animal that tried (and failed) to conquer the skies? Or are both interpretations wrong?

Only new paleontological discoveries will be able to shed light on this matter. In the meantime, however, other fossils have been found which demonstrate that real birds (sometimes still with teeth) were flying immediately after the time of the *Archaeopteryx,* fully in the age of the dinosaurs.

Some scientists even believe that modern birds already existed at the time of the last *Archaeopteryx.* This may be a slight exaggeration, but many fossils suggest that birds did indeed undergo rapid evolution.

For example, some animals, such as the *Sinornis,* which was recently discovered in China and lived "only" 15 million years after the *Archaeopteryx,* already have much less pronounced reptilian features. Shorter tails, legs less suited to running, and stronger flying muscles confirm the evolutionary trend toward modern birds.

One thing is certain: flying above the heads of tyrannosaurs 80 million years ago (on December 24th) were birds similar to those we know today, and many of them belonged to the *Charadriiformes* group, which includes modern seagulls and other shorebirds.

There were also other, more curious birds, which did not leave descendants, such as the *Ichthyornis,* a kind of seagull with teeth; the *Gobipteryx,* which lived in the Mongolian desert; and the *Hesperornis,* a toothed fish-eating marine bird, which later (like penguins) lost the ability to fly.

The emergence of feathers

The *Archaeopteryx* is still standing motionless, its open mouth lined with tiny, razor-sharp white teeth. The movements of its tongue betray short, rapid breaths. It is clearly trying to cool down by dispersing heat from the mucous membrane in its mouth.

This is the price it has to pay for a down jacket in a tropical climate like this. But on the other hand, feathers (like fur) are excellent for keeping heat in, and this characteristic is extremely useful for a reptile like *Archaeopteryx* which needs a lot of energy for flight.

How did feathers develop? It is not hard to imagine their origin. If you lengthen the scale of a reptile's skin you obtain a structure that (like that of the cassowary) retains heat. If this structure then branches out and forms a fringe, creating a velcro-like mechanism to keep the fringes together, the result is a feather. Not only is this feather replaceable, it is also branch-resistant; that is, the wing of a bird (unlike that of a bat) is not damaged if punctured by a branch.

A very ingenious solution indeed, even if the main question still remains: why and how did flight evolve?

Some feel that the flight of the *Archaeopteryx* arose from simply jumping from tree to tree, first gliding and then gradually developing more and more sophisticated and active flight involving the beating of wings. The advantages were that it could catch insects and small prey that were inaccessible to other animals and that it could stay off the ground and out of the range of most predators.

Others, however, believe that it developed from hopping and gliding (rather like the famous Road Runner cartoon character). When faced with an unpleasant encounter, it would disappear from the predator's sight for a moment with a quick jump and short flight. This tactic is still used by such diverse animals as quail, grasshoppers, and flying fish.

It must be emphasized that, contrary to popular opinion, the *Archaeopteryx* was able to beat its wings. Its flight was, therefore, active to some extent, even though it probably did include a lot of gliding.

The *Archaeopteryx* has finally stirred. With quick, short steps, it moves to the edge of the channel beside the islet. Then, suddenly, after a short run, it spreads its colorful wings and takes to the air, soaring over the water and disappearing behind some tree ferns.

But other kinds of wings can also been seen in the skies of the Jurassic: completely different wings, lacking both quills and feathers.

Flying reptiles

Moving to another part of the archipelago, we come to a place where the muddy shores of the channels and lagoons are covered with *Compsognathus* tracks. Underwater, crabs and shrimp can be seen scampering in all directions while an endless variety of fish move almost imperceptibly below the surface.

Suddenly an animal darts forward. We barely have the time to make out that it is a flying animal, something halfway between a lizard and a bat. Then we watch as it skims over the surface of the water and glides away with a tiny fish in its jaws.

It is about the size of a seagull and has the same agility, but it is certainly not a bird. Protruding from between its hind legs

"The *Rhamphorhynchus* skims over the water, its snout just above the surface, and then glides away with a tiny fish in its jaws."

is a long tail, faintly reminiscent of that of the devil ending in a flat, diamond-shaped spearhead, or aileron, to stabilize flight. Its wings, similar to those of a bat but much wider, are made of a membrane that extends from the front to the hind limbs and is stretched like a kite by a long finger. Only three fingers of the hands are left free, but they are equipped with strong claws. The tapered snout is set with a series of long, sharp teeth that are pointed forward like spears and are very effective for catching fish.

Reaching an outcrop on the other shore, the flying reptile lands heavily. Its wings shrivel and fold like an umbrella. This is the first pterosaur we have come across, a *Rhamphorhynchus*. On land, it looks just like a lizard with two exceptionally long fingers.

Pterosaurs were another attempt made by reptiles to conquer the skies, an attempt quite distinct from that of birds. In fact, pterosaurs, unlike the *Archaeopteryx*, are only distantly related to dinosaurs. They form a group of their own, like that of crocodiles. Therefore, they are not "flying dinosaurs" as is commonly believed, but are derived from an earlier evolutionary branching.

Fossil findings provide no help in reconstructing their history, but it is reasonable to assume that at a certain point a reptile similar to a lizard adapted to living in trees and developed a membrane stretched between its front and hind limbs. In this way, it could "jump" from one branch and slowly "fall" to another. This reptile (the fossil remains of which are still being sought) must have lived between 255 and 245 million years ago. Its fourth finger (the ring finger) was probably already longer than the rest to keep the flying membrane taut (in pterosaurs, the little finger disappeared altogether).

Therefore, the line was quite distinct from that of the *Icarosaurus*, which we saw at the beginning of the last chapter and which simply had a membrane along the sides of its body, as some squirrels and arboreal marsupials have today.

The next stage must have been to improve on these characteristics and, above all, to develop the ability to beat their wings, allowing for the transition from gliding or "falling" flight to active flight.

Prehistoric fighter planes

Now the *Rhamphorhynchus* is no longer alone; other pterosaurs have taken to the air. Some are flying just above the water, others over the land. They have a short tail and a long tapered nose with forward-pointed teeth.

These are pterodactyls, smaller and more able fliers. Some manage to catch insects in flight; others harpoon fish. Although pterodactyls may look like a simple variation on the *Rhamphorhynchus*, they actually form a group of their own. In fact, there are two types of pterosaurs: those with long tails (*Rhamphorhynchoidea*) and those with short tails (*Pterodactyloidea*). The former (and more ancient) will become extinct before long. The latter, on the other hand, will witness the entire era of the dinosaurs before they decline, perhaps as a result of competition from birds.

The second group is the one that will develop such odd

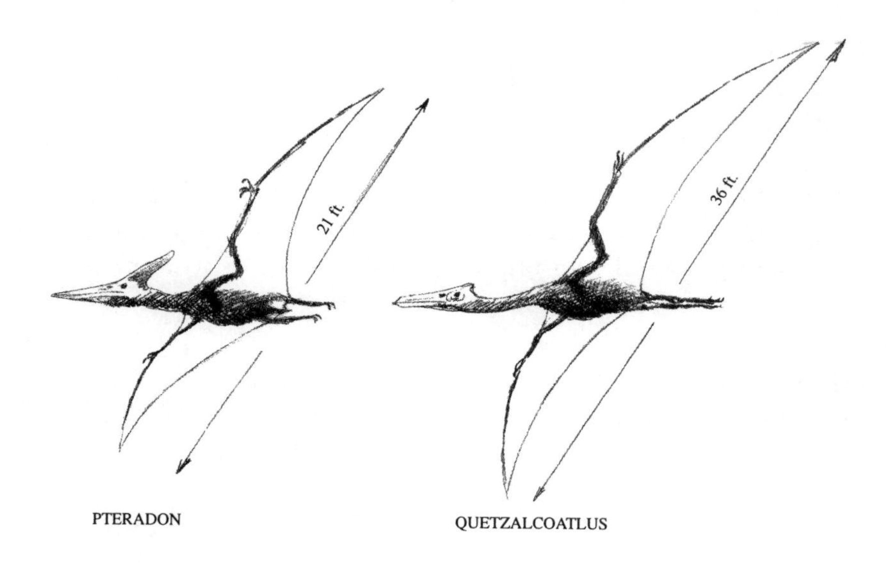

PTERADON QUETZALCOATLUS

forms as the *Pteranodon* (with its hammer-shaped skull) and the *Tropeognathus,* which has keels on either side of its beak to help it "break" the water and fish. It will also generate giant forms like the *Quetzalcoatlus,* a pterosaur with a wingspan comparable to that of a fighter plane.

Some flying reptiles of the *Pterodactyloidea* group have alighted on a rocky outcropping overlooking the water. They look very unusual: they resemble pterodactyls in size and shape but have completely different teeth, which are so long that they sprout like the hairs of a beard from the sides of their jaws. These are *Ctenochasma,* literally "comb-jaw."

What can so many (over 250) long, thin teeth possibly be used for? A *Ctenochasma* stalks into the water, dips its beak below the surface, and moves it back and forth. It is obviously capturing small invertebrates (crustaceans, larvae, and possibly small fish), using the teeth of its "comb" as a filter and net. It is surprising how similar this method is, in principle, to the one used by flamingos today: they also draw in water, shaking their beaks and then capturing crustaceans and other tiny animals in their filaments and lamellae.

Did dinosaurs live in trees?

After everything we have seen and heard about flight, one final question automatically comes to mind: did dinosaurs live in trees?

If the *Archaeopteryx* and other animals started to fly by jumping off branches, their nonflying ancestors must have lived in trees. If we look at reptiles today, it is clear that many—snakes, for example, but also various kinds of lizards—climb (and sometimes even live in) trees.

Therefore, it seems entirely plausible that some herbivorous or insect-eating dinosaurs may have adapted to arboreal life. These would have been small animals, of course. But given the frequent evolutionary "convergences" between different animals, it seems perfectly reasonable to assume that there were arboreal dinosaurs that occupied ecological niches analogous to those of lemurs or monkeys today.

Fossil remains of this kind have not been found, but little effort has been put into the search for small dinosaurs. To date, paleontologists have dedicated most of their attention to the discovery of larger animals, and that is only natural. Not only are they more spectacular, they are also easier to find and to interpret. The smaller forms are rarer because their small bones are more difficult to preserve in general, particularly if they lived in forests where fossilization is almost impossible. Consequently, little is known about them.

It is likely that the most sensational surprises in the future regarding the dinosaur world will come from this type of fossil. And we should not be too amazed if arboreal forms of dinosaurs are discovered. Evolution often follows the same "script," simply recasting it with different actors.

"In a few seconds the ferocious deinonychusians are upon the *Tenontosaurus* in huge leaps."

15

The Rulers of the Cretaceous

DECEMBER 19TH (135 MILLION YEARS AGO)

A jaunt to the poles

We have reached the South Pole and have not had to put our snowshoes on once. In fact, there is neither snow nor ice here, only forests, lakes, rivers full of fish, and swarms of insects. This is precisely why we have come to the Antarctic: to examine this strange phenomenon. The climate and the landscape are completely different from those we know today. No one, given what we see now, would ever guess that this place will be frozen over one day (with record temperatures of –130 degrees Fahrenheit).

The mild climate, perfectly suited to plant and animal life, is surely an anomaly. Only the most oblique rays of the sun reach this part of the earth. The continent should actually be covered by an ice pack. But the unusually mild climate could be the result of a minor "greenhouse effect" caused by the high concentration of carbon dioxide in the atmosphere (this is the hypothesis put forward by Canadian paleontologist Dale Russell, curator of fossil vertebrates at the National Museum of Natural Sciences, Ottawa, and an internationally known expert on dinosaurs).

There are other anomalies as well: the extreme inclination of the sun (which never rises more than halfway between the horizon and its zenith during summer) produces a number of strange effects on the vegetation.

While the competition for the sun's rays (that is, the attempt to grow higher than one's neighbor, spread one's foliage, and steal the sunlight) is "vertical" in the tropical rain forest, competition in this region is almost "horizontal." In order to obtain as much sunlight as possible, a plant has to be *in front of* its neighbor (rather like sailboats in a regatta, which cut each other off and "steal" the wind).

This leads to very surprising consequences for forest growth. Given the low slanting rays, the most favorable position for germination of a seed is in front of the others so that it can get as much light as possible. Any seed that falls into the shade of its parent has little chance of survival. As a result, younger generations tend to grow on the outer edges of the forest, while the trees at the center die of a lack of light. Thus, forests tend to take the shape of a continually expanding ring, with higher and more mature trees on the inside encompassing a central empty area covered with dead tree trunks overgrown with moss and low, hardy shrubs. Seen from above, these forests form circles, ovals, and oblongs that intersect and change shape in keeping with the contours of the terrain.

But what happens to plant and animal life during the long, dark winter months? The growth rings inside tree trunks indicate that the trees stop growing, lose their leaves, and enter a state of hibernation.

Animals probably migrate. Once their food has disappeared,

the large herbivores—at least those living in the outlying areas of Antarctica—may make for greener pastures, followed by carnivores.

The only animals that remain are the ones that are too small to undertake such long journeys—animals such as the upsilophodont, whose fossil remains have been found close to the Antarctic. How do these small animals survive? Some of them may adopt a strategy will known to us: they dig a hole and hibernate.

DECEMBER 20TH (125 MILLION YEARS AGO)

The attack of the Deinonychus

The Jurassic period has been over for about 20 million years. The era of the giant sauropods, stegosaurs, allosaurs, ceratosaurs, and ramphorhynchians is gone forever. A gradual climatic change has altered environmental conditions, causing another major crisis for plants and animals.

The climate is still tropical, but it has gradually become drier. It seems that these climatic changes were responsible for the extinction of the dinosaurs, even if the mechanisms and the initial causes for the process are still not clear.

We are not far from the equator and the landscape around us is much like the modern savannah, with small "islands" of dense plant growth. Vast expanses of soil laid bare by the lack of vegetation attest to the hot, dry climate. A slight breeze eddies silently over these red scars, picking up pieces of fern and tossing them about..

In the distance, almost on the horizon, the mirage of a huge lake created by the heat waves reflects the trembling figures of three bipedal dinosaurs on the run, probably in search of prey.

Hidden in the vegetation, a large reptile observes the distant figures. This herbivorous dinosaur, the *Tenontosaurus*, looks like a large cow with an elongated neck and a very long tail held parallel to the ground.

Paleontologists have not yet explained the long, flat tail. It may have been used as a whip or as a solar panel to accumulate heat in the morning sun. In fact, this dinosaur has no effective defensive weapons and its long claws are more suited for pulling up plants and walking than for defense. This makes this 21-foot beast, which prefers to be on its own, perfect prey for carnivores.

But it is difficult to catch it off guard: its eyes, nostrils, and "ears" (actually auditory holes) constantly monitor the surroundings, on the alert. In addition, its body is well camouflaged with the colors of this tree-covered savannah.

The tenontosaur suddenly raises its head. Something has caught its attention. Its eyes see nothing out of the ordinary, only the slow swaying of branches in the breeze, and yet, the animal is obviously uneasy; it lifts its head and sniffs the air.

A distinct sound now carries on the air: that of reeds being crushed underfoot. The tenontosaur realizes the danger, but too late. Before it can move, a number of shapes shoot out from the underbrush, giving off hideous sounds. They look like a pack of ferocious dogs.

This is a group of *Deinonychus*, hungry carnivores the size of a human. Armed with razor-sharp teeth, they also have a huge sickle-shaped nail on each hind limb with which to rip open prey. While moving, they keep their heads lowered and tails outstretched to balance their forward-projected bodies. The sun reflects off the well-honed knives in their half-open mouths. Each movement is a declaration of speed and power: these are killers, consummate and merciless.

In a few leaps they are upon the tenontosaur. Each *Deinony-*

chus attacks a different part of the body, sinking its jaws and claws into the victim's flesh. With a quick backward flick of the head, their saw-edged teeth tear away ample portions of skin and muscle, leaving deep gashes. The tenontosaur wavers, tries to take cover, and with a well-aimed blow of its tail crushes the chest of one of its assailants. But ribbons of blood now stream down all over its skin.

As if excited by the sight of blood, the *Deinonychus* dig their teeth in even further and tear open new wounds with their lethal hind nails. Two of them have gashed the tenontosaur's belly and are starting to snap at the gushing internal organs.

The tenontosaur manages to smash another assailant with its tail, but the battle has been lost. Weakened by the loss of blood, it leans against a tree and slowly sinks to the ground. Perhaps it no longer even feels the pain as darkness first clouds its eyes and then overcomes it.

The beasts scramble over the corpse—which is shaken by a few final tremors—biting at the vital points and tearing away the flesh.

The hunt is over. It now gives way to hissing and aggressive behaviors used to establish the group hierarchy for the meal. Movements are rapid, almost abrupt; even when eating the prey, they swallow savagely without chewing.

Silhouetted against the low-hanging clouds of dusk, two pteranodons glide majestically overhead. They finally alight on an escarpment where they have probably built their nests. All sounds gradually die down as the breeze caresses the darkening savannah, bringing another day 125 million years ago to an end.

DECEMBER 22ND (95 MILLION YEARS AGO)

A swim in the seas of the Cretaceous

The rocks we are climbing down are black and slippery. Behind them lie steep cliffs populated by an incredible number of flying creatures. We can see "seagulls with teeth" (*Ichthyornis*), real birds, and many pteranodons (which are in a certain sense the reptilian version of pelicans, albatrosses, and frigate birds).

The flight of these enormous pterosaurs is majestic. Buoyed by air currents, they sail over the waves like gliders. Then they plunge head-first into the sea, only to regain elevation immediately afterwards, beating their wings while firmly holding the prey in their beaks. Some have a wingspan of up to twenty-seven feet.

This spectacular scene unfolds before us as we dive into the seas of the Cretaceous. The fish that we see underwater are different from modern ones: they all tend to be very long (like barracudas or herrings) and have large scales. There are almost no "round" fish, like the ones found in tropical aquariums, even though these seas have boundless coral reefs.

Ammonites of all sizes float around us like tiny hot air balloons. Their spiral shells oscillate rhythmically as each "squirt" of water from the muscular "funnel" used to pump water propels them forward. They have a funny kind of flat, protective "hat" on their heads and look warily at us with their fixed, glassy eyes.

As we move away from the coast, the bottom sinks away and the water becomes a more intense blue. We are not totally at ease about swimming in such an immense and unknown environment where predators could appear at any time.

In fact, these seas are inhabited by mosasaurs: very aggressive and voracious reptiles that look like eighteen-foot-long crocodiles or water monitors.

Extraordinary fossil remains document that ichthyosaurs were viviparous and that offspring were born tail-first. Dolphins today are born in the same way, since they have to rise to the surface to breathe immediately after birth.

There seems to be some kind of commotion near the surface to our left.

Sharks and ichthyosaurs

We approach with caution: some sharks are haggling over a large prey. Five or six of them are busy tearing it to pieces in a maelstrom of water and blood; the prey—torn to tatters—is no longer recognizable. Sharks have become the absolute rulers of these oceans, ousting the reptiles (ichthyosaurs) that controlled them for over 100 million years.

One of the last ichthyosaurs approaches the sharks' banquet. It is almost identical to a dolphin: six feet long, with a long snout, small cone-shaped teeth, a stabilizing dorsal fin, and smooth skin. Unlike a dolphin, though, it has a vertical tail (like a fish) rather than a horizontal tail and a pair of long hind fins.

It is amazing to see how similar habits have resulted in similar shapes, to the point that, although a reptile, the ichthyosaur no longer lays eggs and gives birth to its young in the water. This is one of the clearest examples of the way in which two animals of completely different origin can end up resembling each other if they occupy the same niche.

This probably happens because there is not an infinite number of solutions to the same problem. In the end, the sum of similar strategies for dealing with identical problems (from hydrodynamic shape and maintaining body heat to the kind of teeth needed to hunt certain kinds of fish) produces almost identical animals. These are the so-called evolutionary convergences we have already discussed.

Opposite page: "A mosasaur, a very aggressive and voracious eighteen-foot aquatic reptile, snaps up the ammonites floating around like tiny hot-air balloons."

After snapping up the few unnoticed pieces of meat sinking from the sharks' banquet (an ichthyosaur eats small and medium-sized fish and would hardly bite into a carcass this size), the dolphin-reptile slowly swims away into the deep.

We are witnessing the decline of a group of reptiles which, at its peak, had developed numerous forms (some up to forty-five feet long) and spread throughout the world's oceans. Now it is disappearing forever.

Other great reptiles also developed in these seas, such as the *Kronosaurus,* which reached its prime in the lower Cretaceous. One is swimming by right now. It is terrifying to see.

The *Kronosaurus* can reach up to thirty-five feet in length, but its long, massive skull alone, armed with formidable conic teeth, can measure up to nine feet. This animal looks like a gigantic lizard with its huge head, short tail, and paws transformed into paddles. It has a stocky body with practically no neck. The bulldog of the seas, it is a predator with habits very similar to those of the modern-day killer whale (orca).

But now a very strange "cousin," the *Elasmosaurus,* appears in the distance. It resembles a snake that has swallowed a turtle: it has paws that have been transformed into paddles, a heavy flat body, and a long neck with a tiny head and very sharp teeth.

"A flower! This is the first flower we have seen. It looks like a magnolia and is strongly scented . . . insects, especially beetles, are busy pollinating, but butterflies and bees are still absent. In the background, two *Maiasaurus* are unaware of the host of tiny insects around them which will be active for millions of years after the dinosaurs have gone into extinction."

The elasmosaur swims like a turtle, using its hind legs for propulsion, while its head is stretched forward in search of prey (small fish). Suddenly it stops, its long neck cranes around sinuously and the head disappears into a rocky crag. A few seconds later, it reappears with a small fish in its mouth, which it gulps down in one mouthful before continuing calmly on its way.

DECEMBER 23RD (85 MILLION YEARS AGO)

The perfume of a flower

A flower! This is the first flower we have seen. It looks like a magnolia and is strongly scented. For millions, no, billions of years, the earth has lacked the colors and perfumes of flowers. But now the revolution of the angiosperms, that is, the flowering plants (from the agave, rose, and cactus to wheat, elms, clover, rhubarb, and so on) that can be found throughout the modern world, is in full swing.

This most important innovation was achieved gradually after the appearance of dicotyledons (which include such trees as the beech, birch, hazelnut, fig, and laurel) and monocotyledons (flowering plants with parallel-nerved leaves such as bamboo, rushes, lilies, tulips, pineapple, orchids, and palms).

These plants are the result of another climatic change; in fact, that is now quite different from the temperature in the Cretaceous. It is slightly colder, but above all less stable, with more marked seasonal changes.

Differentiation seems to be the key word at this stage of the dinosaur era. In addition to changing seasons, the land masses are also differentiating: their separation has continued and they are now in positions similar to those of today. The dinosaurs, increasingly separated geographically, have started to specialize: some forms are becoming typical of certain areas and not of others.

What is most striking about the forest we are in right now is its familiar appearance: there are sequoias with their red trunks, cypress trees in the more humid areas, myrtle, boxwood, and heather. Different kinds of ivy climb the tree trunks and "garlands" of mistletoe hang from the branches.

Furthermore, the trees are not always vertical; they are often bent and contorted with low-hanging branches. This is definitely a "modern" world, and a scented one as well. Many trees make quite a show of their flowers. This is the first time in the history of life on Earth that perfumes waft through the air; till now the only scents were those of resins and mosses.

Clouds of pollinating insects swarm around the flowers. They include crickets, dragonflies, and various kinds of beetles, but butterflies and bees are still missing.

The only social insects that are already widespread are the termites that devour the huge trunks of the evergreen trees. Ants have not come on the scene yet either.

This half-modern, half-prehistoric world is well documented by the fossil record. Miraculously, the strata dating back to the Cretaceous have even preserved some flowers intact: charred by fire, they acquired the hardness required to stand up to sedimentation. The fossil flower is known as the *Archaeoanthus*. It is six inches in diameter and looks much like a magnolia blossom.

The outburst of the angiosperms probably started quietly in the early Cretaceous with low vegetation that had flat leaves (rather than needles) and was obliged to grow in marginal areas exposed to herbivores and erosion. But these plants had some extraordinary abilities, the most important being rapid growth, even after mutilation by herbivores. They soon reached the size of shrubs and finally trees, creating small forests along rivers and streams.

DECEMBER 24TH (75 MILLION YEARS AGO)

The intelligent dinosaur

One question has been on our minds since we started observing the great variety of dinosaurs in the Cretaceous. The forms of life are certainly smaller now, even though giants like the tyrannosaur, the triceratops, and various kinds of hadrosaurs still exist. But are there no evolutionary trends in totally different directions, for example, toward greater intelligence?

In the course of our journey, we have witnessed the remarkable development of the nervous system: starting out from a simple "net" of nerves in the jellyfish, it became a "tangle" in the Cambrian marine invertebrates, later an embryonic brain with almost exclusively primary functions, and later still an increasingly evolved organ.

A group hunting strategy like the one put into practice by the *Deinonychus* calls for brain functions of some complexity. But the question remains: after over 150 million years of evolution, are there any dinosaurs in the Cretaceous that show signs of intelligence? If there had not been a catastrophe at the end of the period (causing mass extinctions), would the dinosaurs have given rise to an "intelligent" line?

After waiting for some time up in a tree, we finally spy some *Troodon*. These are dinosaurs with very special and intriguing characteristics. Not far from us, two are searching for food along a riverbank.

Troodon are bipedal animals. As tall as human beings, they have a brain that is much larger than that of any reptile that has ever existed: even larger than the mammals of their time.

It is amazing to see the agility and grace with which their three-fingered prehensile hand turns over rocks and bits of tree

trunk so that their eyes—capable of frontal vision—can seek out prey.

Strangely enough, they seem to have characteristics reminiscent more of apes than of the mammals living at the end of the Cretaceous (which will give rise to apes much later). These are definitely very evolved dinosaurs, able to invent new strategies to deal with new problems.

Their characteristics seem almost "human." Of course, their reptilian nature prevents any kind of transformation in this sense; but can we really exclude the possibility that these forms might have turned into increasingly intelligent animals if they had been given enough time to evolve, say tens of millions of years? Perhaps into the reptilian equivalent of a hominid? Not into a scaly-skinned *sapiens,* but at least into a form substantially equivalent to an *Australopithecus,* or "ape man" (our ancestor from four million years back)? The *Troodon* already seems to possess certain anatomical characteristics suggesting such development.

Of course, a reptile is not a mammal, especially as regards the structure of the brain, and that is the main difference. Could the dinosaurs have solved this problem if they had been given the chance to develop further?

According to paleontologist Dale Russell, perhaps yes. He feels that dinosaurs are very special reptiles and he would not rule out that animals such as the *Troodon* could have evolved into forms with superior intelligence if they had been given enough time.

This hypothesis is very controversial and, of course, cannot be proven. As we all know, dinosaurs were *not* given enough time. A drastic climatic change wiped out a large number of living forms, including the last dinosaurs.

The great extinction of the Cretaceous

Many theories have been put forward to explain this mass extinction (more than eighty are known, most of which are unfeasible or utterly outrageous). Some scientists even feel that the extinction was not sudden, that it took millions of years. This is based on the fact that some dinosaur datings are subsequent to the fatal date of 65 million years ago. But this hypothesis is also challenged, as a number of the datings in question are uncertain: the sediments containing them may have been interpreted incorrectly.

The asteroid theory seems to have gained currency lately. The hypothesis is that an abrupt climatic change was brought about by the fall to Earth of one or more asteroids, which raised such a cloud of dust and debris that it blocked out the sun.

The asteroid theory has one notable advantage: there is some evidence of it. For example, traces of an immense underwater crater have been found on the Yucatan peninsula and its date of origin corresponds to that of the end of the dinosaurs: 65 million years ago. In certain areas in Mexico, geologists have found evidence of the shock wave produced by the asteroid: various layers of vitreous spherules (generated by the extreme heat of impact), debris, and sediments, which provide a geological "record" of the catastrophic collision.

The crater's diameter (approximately 112 miles) also seems to correspond to theoretical calculations based on the size of the asteroid. In fact, scientists have calculated that the asteroid would have to have had a diameter of at least six miles in order to raise enough dust and debris to block out the sun for a sufficient length of time. These calculations derive from studies carried out by experts in atomic warfare, who ponder the effects of a "nuclear winter," that is, of a sudden and catastrophic cooling of the planet caused by the enormous amount of material hurled into the sky during a nuclear explosion.

Did an asteroid cause something of that kind at the end of the Cretaceous period, setting off a chain reaction: the death of plants, a lowering of the temperature, the death of herbivores and, consequently, of carnivores?

The question remains open, mainly because it is hard to explain why certain forms became extinct while others did not. For example, why did all the dinosaurs (both large and small), all flying reptiles, and all marine reptiles disappear, and not crocodiles, turtles, sharks, or snakes, for example?

No satisfactory answer has yet been found to this question. Perhaps the dinosaurs, the pterosaurs, and the marine reptiles had a special "Achilles' heel" so that the sudden cooling affected them more than others?

Fresh water sharks, for example, became extinct while sea sharks survived. Why? It almost seems as though the extinctions were random, as in an airplane crash in which some passengers survive while others perish. The same happens during bombings or epidemics or in concentration camps. Some species, although decimated, do not disappear. The individuals that survive give rise to new descendants which repopulate their niches.

It is important to underline another aspect of these mass extinctions which is often overlooked: the dinosaurs may have disappeared at the end of the Cretaceous in any case. According to the estimates of some paleontologists, most of the dinosaurs had already gone into extinction in the 10 to 20 million years prior to the catastrophe. At the time of the catastrophe itself, only eight of nineteen families still existed, with only ten genera and a total of twelve species. Marine reptiles were also on the decline.

The debate is still open, but three things may be said in summary: (1) most scientists agree that some climatic crisis took place; (2) there is reasonable evidence to suggest that at least one large asteroid hit the earth during that period; and (3) there is disagreement about the real effects of the asteroid and about the time that it took the dinosaurs to disappear: most paleontologists do not agree with the idea of sudden extinction (of the "nuclear winter" type), and feel that it probably took longer and may have been linked to other kinds of climatic changes.

More will be found out about this in the future.

A Triceratops *stampede*

The *Troodon* suddenly raise their heads to look in a certain direction. They evidently sense something. Even we, perched up high in our tree, can feel a dull vibration which slowly turns into a low rumble, like an approaching herd on the gallop. In one bound, the *Troodon* disappear just as a confusion of horns, beaks, and bodies breaks through the vegetation in a cloud of dust and

The *Triceratops* cross the small river.

stones and pounds into the river, sending up spray on all sides. The herd is made up of tens, maybe hundreds, of stampeding *Triceratops.*

Over the splashes, we can hear the deep "grunts" regularly given off by the *Triceratops.* These animals are quadruped herbivores with habits similar to those of the buffalo. They can reach lengths of up to twenty-seven feet and a weight of ten tons (twice that of an elephant). Two horns are located above the eyes, while a third projects from the nose. A large bony collar fans out at the back of their head.

All these features probably serve in their fights for possession of territory and females. But the large collar may have another use: the dense network of blood vessels on its surface could provide a "solar panel" to heat the animal's blood.

The *Triceratops* jam together as they ford the small river. Moving together provides better protection against attacks from their enemy, the *Tyrannosaurus rex*. Bunched together they form such a mass of muscles, horns, and spikes that a tyrannosaur would have some difficulty in raiding such an armored column.

In fact, a tyrannosaur is unlikely to attack a *Triceratops* (just as a lion is loath to attack a buffalo) unless the victim is alone or, better yet, young. It is much easier to hunt less heavily armored, weaker (and perhaps more "tender") animals such as hadrosaurs.

The herd of *Triceratops* crashes away through the vegetation, trampling everything in its path. In spite of their strength and vitality, these animals are also headed for extinction.

DECEMBER 25TH (65 MILLION YEARS AGO)

Tyrannosaurus rex*: the last great hunt*

The catastrophe is about to strike; the end is near, but no one realizes it. For the dinosaurs, this day is just like all the rest. They spend it grazing, fighting for territorial supremacy, mating, hunting, giving birth—and dying.

It is strange to think that the strength of these giants is nothing compared to the power of a natural event: a change in climate, a sudden cooling—perhaps as the result of an asteroid falling from the sky.

At least twelve other extinctions are known to have taken place before this time. But this is certainly the most impressive because it suddenly wiped out the main characters of a story that lasted approximately 170 million years.

It is to the *Tyrannosaurus rex* and its last great hunt before its demise that we want to dedicate this last day.

It is dawn. Half hidden in the woods, a *Tyrannosaurus rex* is preparing for attack.

This is a large animal, about forty-two feet long and eighteen feet high. Its huge curved and pointed teeth are as big as bananas; they even have special bony supports between them to keep them in place on impact and while biting and tearing. The animal's jaws are like those of a shark: they are not meant for holding the prey, only for tearing its flesh.

The sun has just found its way through the leaves and traces a mottled pattern onto the animal's back. The spots of light seem to dance on the wrinkly skin covered in a mosaic of tiny horny tubercles. But suddenly the spots cease to move. The tyrannosaur has come to an abrupt stop, slowly lowering the one leg that was still raised to the ground.

In the boggy clearing before it are some unmistakable shapes: a group of *Parasaurolophus* feeding greedily. These animals are bipedal plant-eaters. They measure about thirty feet in length and twelve to fifteen feet in height; have strong, three-pointed hooves; and hold their long tail straight over the ground. The most noticeable feature, however, is the head of the male: it faintly resembles that of a horse, but has a long "tube" that runs from the tip of the nose to the back of the head, projecting beyond like a periscope.

Other heads bob up and down in the vegetation.

The tyrannosaur slowly crouches to the ground. Its breath has become short and rapid. It sinks its hook-shaped claws into the soft sand while waiting for the right moment.

The group of *Parasaurolophus* has moved farther out into the open. The deep growls that the males regularly emit signal that all is well—there is no danger.

This is the moment the tyrannosaur has been waiting for. Very, very slowly, it gets to its feet. A slight gurgling escapes its half-open mouth. The mottled light begins to dance once again on its back. Its eyes target its victim.

A male suddenly whirls around, attracted by a glimmer in the shaded foliage. After a moment's hesitation, it senses the attack and lets out a deep roar. But it's too late. The tyrannosaur has sprung forward, its body deformed at every step by the swelling and distension of its muscles.

Its claws, like cleats on a sprinter's track shoes, dig deep into the ground, hurling its body forward. Its jaws gape terrifyingly. All that the herbivores can see is an enormous dagger-framed trap charging toward them.

At a speed of close to twenty miles per hour, the tyrannosaur's seven tons smash into its victim, a female *Parasaurolophus* slowed down by some shrubs.

The initial contact is devastating. Its jaws close like a trap on the female's haunches while its neck bends and its snout deforms slightly to absorb the impact. The tyrannosaur sinks its white teeth into the prey's muscles and with a violent jerk of its head rips away a large portion of the victim's leg.

Thrown to the ground, the *Parasaurolophus* cries out desperately as it makes a weak attempt to rise to its feet. But the tyrannosaur's teeth now sink in between the rib cage and the back. Bones can be heard snapping as the assailant tears away more flesh with another violent jerk of its head.

The flexibility of the tyrannosaur's jaws and joints allows the skull to deform with each bite, flattening out on the sides. Now the *Parasaurolophus* no longer reacts. Lying in a pool of blood, its rapidly weakening breath raises increasingly smaller eddies of dust until it finally just closes its eyes.

For a moment, the tyrannosaur raises its bloody muzzle in

The huge tyrannosaur, eighteen feet tall, observes a group of *Parasaurolophus*.

conquest. The alarmed cries of the other *Parasaurolophus* echo through the woods. But too late; the hunt is over.

The tyrannosaur leans into its feast once again. Four of its teeth, broken and blood-covered, lie on the ground beside its victim. These are its tribute to the hunt, but they will soon be replaced by others that are already in place in its jaws.

In a very short time, however, much more powerful jaws are going to crush this and all other dinosaurs.

The "feathered" dinosaurs

With the great crisis of the Cretaceous, and the sacrifice of the giants, the earth is like a theater that has been emptied. Most of the spectators have left, leaving empty seats in the stalls and the gallery.

But these seats will not remain empty for long. In the corridors and outside, individuals who were left standing or could not get in will take the unexpected opportunity to occupy the new "niches" that were previously inaccessible. These are the small mammals that have been living a marginal, often nocturnal existence for tens of millions of years.

In the next chapter, dedicated to the advent of mammals, we will see the extraordinary change that took place in their world with the rise of totally new—and sometimes very large—forms (some of them will reach the size of the large reptiles).

But before following the mammals through the last stages of evolution preceding that of human beings, let's take a look at some of the last relatives of the dinosaurs, that is, the "feathered" branch that managed to survive the extinctions of the Cretaceous: birds.

As we saw in the last chapter, birds are the closest relatives of the dinosaurs, much closer than flying or marine reptiles. At the time of the great crisis, they were already making their presence felt in the skies, winning out in the competition with many pterodactyls, which were on the decline.

It would be impossible to follow the route which led them,

in the course of a few tens of millions of years (a few days on our calendar), to give rise to the variety of known forms: from the albatross to the finch, the vulture to the hummingbird, the parrot to the flamingo, the eagle to the toucan. But it is interesting to recall some of these forms, in particular, the so-called nonflying birds.

The "Tyrannosaurus bird"

After the disappearance of the dinosaurs, some large birds emerged which were vaguely reminiscent of such light, cursorial dinosaurs as the struthiomimus. One example is the *Diatryma,* similar to the ostriches but sturdier. This bird, which appeared in Europe and North America during the Paleocene (around 50 million years ago) was about six feet high, had practically no wings, terrible claws, and a head as large as a horse, with a very strong beak for tearing meat.

The *Diatryma* could be considered a kind of tyrannosaur-bird. Paleontological reconstructions based on skeletons show that it was a terrible predator, even if not much is known about its habits. When the profile of its skeleton (adding an imaginary tail) is compared with that of a young carnivorous dinosaur, the resemblance is striking.

This group of great nonflying bipeds has continued to be a part of terrestrial fauna, although it has taken different evolutionary routes (producing diverse animals): about 20 million years ago, another giant predator appeared in South America—the *Phorusrhacus.* It also had an enormous skull about one and a half feet wide and a terrible beak. Cousins of this frightful predator can still be seen in marshes, along rivers, and on the roofs of houses today: they are cranes, coots, water hens, and bustards.

"*Diatryma* was almost a kind of tyrannosaur-bird."

Another, completely different line of large nonflying birds which started in the Cretaceous is that of the ratites, i.e., flightless, running birds, which was to lead to the modern ostrich, nandu, emu, and cassowary. Until three hundred years ago, the *Aepyornis,* which was almost nine feet tall and weighed up to half a ton, lived in Madagascar.

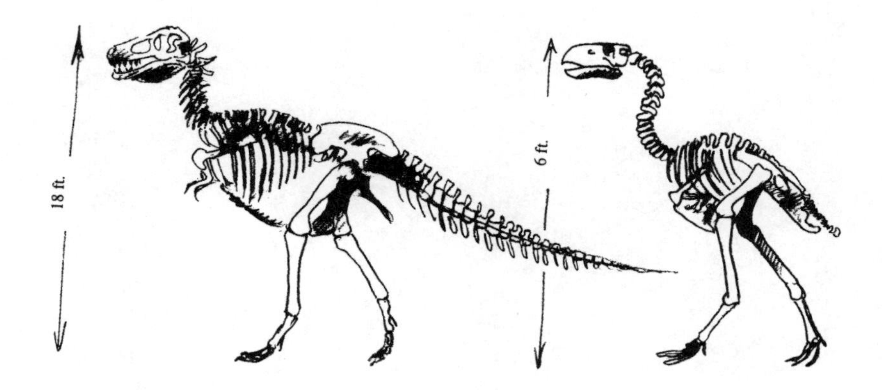

Comparison of the skeletons of *Tyrannosaurus* and *Diatryma*. The similarity is quite evident, apart from the size.

Even taller—it could reach up to ten and a half feet—was the moa (*Dinornis maximus*), the greatest bird ever to walk the planet. It still existed less then two centuries ago in New Zealand before being wiped out by the Maori.

Of course, the moa, the *Aepyornis*, and even more so the ostrich and the cassowary (not to speak of the eagle, crane, lark, and heron) are lines which are distinct from those of the dinosaurs: 100 or 200 million years of evolution make quite a difference. But an increasing amount of research has demonstrated their kinship (which was often overlooked) with the great reptiles of the past.

A kinship which has induced one paleontologist to utter the rather provocative quip: "Who said dinosaurs are extinct? They are still with us. They wake us every morning with their song; they are birds."

It is dusk. A *Ptilodus,* about twenty inches long, moves agilely on the upper branches.

16

An Explosion of Mammals

DECEMBER 26TH (63 MILLION YEARS AGO)

A surprising squirrel

Returning after the Great Extinction is a bit of a surprise: "only" two million years have gone by since the catastrophe eliminated the dinosaurs and many other species, but apparently a few traces remain. Plant life has returned and the climatic scars have healed, but the planet looks half-deserted.

Walking along riverbanks and marshes, we have searched vainly for some of the great reptiles of the past. The classic watering holes, the lakes, the fern fields are now barren.

It is dusk. Leaning against a tree, we look up to watch an animal scamper from branch to branch. It looks something like a squirrel: it's approximately twenty inches long, has eyes set in the front of its head, and moves with great agility.

Now it has stopped to gnaw at something, perhaps seeds or

some nuts. We can see its large incisors, resembling those of rodents today, but longer and more curved. This is a *Ptilodus,* a primitive mammal that lives in trees and makes brief raids on the ground.

The vegetation is now much like that found on Earth today, with broad-leafed trees and other plants similar to those in our woods and, above all, an abundance of flowers. A shrub not far from us is a veritable explosion of large red, strongly scented blossoms, and it has attracted a host of dancing insects to pollinate it.

Turning back to observe the squirrel, we almost step on a long, slender form that slithers away through the grass: a snake. It is large in size, and dark in color, and looks like a viper. In a second it's gone.

Suddenly we realize why the *Ptilodus* is hesitating to come down from the tree. Living on the ground in this area must not be very pleasant. Large snakes and other predators infesting the area are always on the hunt for small mammals (and especially their young). It's easy to understand why it is safer for these animals to live and nest in more inaccessible places (even if many snakes can climb trees).

The *Ptilodus* does not move a muscle. Ever since it detected the presence of the snake, it has remained poised, obviously in a state of alarm, ready to act, raising its head only ever so slightly.

Then again, all of its ancestors had learned how to live with danger: reptiles have been their enemies for millions of years. Indeed, the *Ptilodus* and all the other small mammals were forced to occupy marginal environments during the era of the dinosaurs. If anything, the situation seems to be changing now after the disappearance of the large reptiles.

Today, paleontologists are trying to reconstruct the history of these first mammals. It is an intricate story, full of evolutionary branchings, which unfolded over an extremely long period of time, almost as long as that of the dinosaurs.

But reconstruction is also complicated by the fact that the bones of these small and fragile animals, unlike those of dinosaurs, have largely been lost. Moreover, in the past, paleontologists have generally been more attracted by the remains of the great reptiles than by the minuscule fragments of these mammals.

A clandestine life

However fragmentary, fossil remains have nevertheless allowed us to understand that these animals lived hidden away in their burrows or dens, ate insects, and were very small. For example, the *Morganucodon*, which lived in the upper Triassic, was only four inches long (shorter than the scorpions of the time) and very similar to a small rodent. It was about the size of a mouse, even though its skeleton was different.

Some may even have been meat-eaters. One such animal is the *Sinodonodon*, which also originated in the Triassic and had differentiated teeth. Teeth, as we know, are the parts of the body best preserved by fossilization. The names of many of these small mammals end in *-odon* (Greek for "tooth") because of the characteristics of their teeth. Sometimes, a piece of jaw or a tooth or two is all that has been found of the entire animal, but teeth have the advantage of revealing many things, usually far more than any other part of the skeleton (they indicate the kind of food eaten and, therefore, certain related behaviors, the evolutionary group of origin, and so on).

The only fossil remains that have been found of the *Purgatorius*, the small mammal resembling a tree shrew and believed to be one of the ancestors of primates (and therefore human beings) is a piece of jaw with a few teeth. But that is sufficient for an approximate reconstruction of some of its characteristics. The *Purgatorius* was, for example, very small: scientists calculate that it weighed no more than twenty grams, about the same as a pack of cigarettes.

The *Purgatorius* probably still lived in similar environments not very long ago; it is on trees like this that it started its evolutionary career. We have been keeping an eye out for it, but have not been able to spot one.

But there were also much larger and more aggressive ancient mammals. One was the *Triconodon,* which lived in the Jurassic and whose remains have been found in Britain. It was the size of a large cat and was a carnivorous predator.

A precipitous descent

The *Ptilodus* has finally moved: it has reached the trunk and is now making its way down toward the lower branches. It descends just like a squirrel, that is, precipitously, digging its long claws deep into the bark with the help of a special joint in the ankle which allows it to swivel its digits around toward the back. Dashing out onto a new branch, it wraps its long tail around it and, like a spider monkey, swings down to a lower limb to reach some succulent nuts hidden among the leaves.

It is surprising to see this small mammal in action. It seems to combine the features typical of a squirrel, a mouse, a monkey and a lemur: it basically looks like a "modern" animal. But its evolutionary line (that of the *Prototheria*) is destined to die out, giving way to other, similar animals that will live in the same environment and will, therefore, by convergence, take on similar characteristics and habits.

The time has come for these small mammals. The disappearance of the dinosaurs has not only opened up new environmental niches and new space, but has also allowed them to do something which they were unable to do for millions of years: grow in size.

On the grasslands, in the forests, on the savannah, and in the seas, mammals have undertaken a new kind of development, one that will lead to increasing size (sometimes gigantic) and increasing complexity, making them the new rulers of the planet.

It's easy to call them mammals

We have overlooked one minor detail concerning our small prehistoric squirrel, the *Ptilodus*: it is a mammal, but a special kind of mammal since it belongs to the same evolutionary line (the *Prototheria*) as the monotremes, that is, egg-laying mammals.

It may be surprising to find out that the earliest mammals laid and hatched eggs. But then, they descended from mammalian reptiles and the transition from the egg to the placenta must have taken some time and involved a number of steps.

No traces of the various "links" of this chain of events have been handed down to us, but in the course of 190 million years, a large number of mammals appeared and disappeared. Those that still exist on the earth today can give only a very vague idea of past "models." Nevertheless, some extant animals may help us to understand this transition. There are, in fact, four contemporary mammals that seem to summarize this evolution from egg to placenta: the platypus, the echidna, the kangaroo, and the antelope (or any other placental mammal).

Four animals, four stories

The platypus, an Australian monotreme, is a mammal that lays and hatches eggs exactly like a reptile, but also has mammary—milk-producing—glands on its abdomen.

The echidna, another survivor of monotreme fauna still found in Australia, resembles a hedgehog with a long beak-shaped nose and thick fur. Not only does it nurse its young, but it also has a "pouch" on its abdomen in which it incubates its eggs, while continuing its daily activity, until they hatch.

We all know the kangaroo, a marsupial that also inhabits Australia: its young start to develop inside the placenta but are born while still almost embryos. They then climb into the mother's pouch where they suckle and complete maturation.

The antelope is a placental mammal like the mouse or the human. The young develop entirely inside the placenta and are fully developed at birth. They are then nursed by the mother.

Four animals, four reproductive processes, four evolutionary stories. The links between these various "models" of mammals are difficult to establish, but they may enable us to understand how some transformations could have taken place and then evolved in certain common ancestors or lines.

During the last 65 million years, from the disappearance of dinosaurs to the modern era, the history of mammals has been full of changes and developments that have involved much more than the reproductive sphere. Strange and surprising animals have appeared, some of which have given rise to modern animals.

Not all the tracks of this immense evolutionary "railway network" can be followed. What we will try to do in this chapter, then, is to provide brief sketches of the animals that have somehow characterized their time or that deserve to be remembered for some special trait. A series of "newsreels," shot at different times and on different continents, will allow us to observe—at close range—some of the protagonists of the most recent stages of evolution: small, large, and even very large animals, all descendants of the "mice" and "squirrels" of the Cretaceous.

DECEMBER 27TH (50 MILLION YEARS AGO)

Uintatheria at the lake

A deep, blue lake is set in a lush forest. Although it resembles the great African lake region, this area is located in the North America prairies, close to the Rocky Mountains.

The vegetation is tropical. Flourishing palms and breadfruit trees brim with songbirds and various new life forms. The empty theater has been filled by a myriad of small mammals that have been transformed in a very short time.

Before us, on the shores of the lake, two huge animals are rolling in the mud. From afar, they look like rhinoceroses, but they actually belong to a group whose origins are not yet clear: the *uintatheria* (these are called *Eobasileus*). Over ten feet long and six feet high, they are powerfully built, with large, heavy legs and strong backs. But most impressive of all are their heads: long and flat, they have two enormous canines sticking out from their upper lip and a battery of six monstrous skin-covered horns (like those of a giraffe) protruding from their skull. The horns look like huge nails hammered into their nose, eyebrows, and forehead.

It is amazing to think that tiny "mice" and "squirrels" could have turned into giants like this in such a short time (only one and a half days on our calendar). These animals now occupy the lands and eat the rich grass of their former—now extinct—rulers, the dinosaurs.

A few uintatheria are grazing, using their tongues to help them tear out the plants. Others are trotting along the shore of the lake, sending up a shower of spray. Finally, we can get a better view of their monstrous heads. They have a terribly stupid look about them; not a glimmer of intelligence projects from those tiny eyes tucked away under the horns. The amount of muscle and bone may have increased, but the brain must still be very small.

Study of the skulls of these animals has confirmed that they had a very small brain volume. But then, their lifestyle and environment were not very stimulating (there was little competition for food and few predators) and required little intelligence.

The *Coryphodon*, which inhabited these regions before the uintatherium (and which resembled a hippopotamus or a large tapir) was even less cerebrally endowed: paleontologists have calculated that half a ton of body weight corresponded to roughly 90 grams of brain.

The group of uintatheria gradually moves away from the lake. Some of the animals are partially covered in mud: this is a way of keeping their skin moist—a method still used by hippopotami and elephants today.

The interesting thing is that this kind of environment will eventually shape the animals succeeding the uintatheria after their extinction (approximately 35 million years ago) in the same way. In fact, the brontothere, which "started out" 50 million years ago resembling a horse, converged toward the shape of a rhinoceros, even though it came from a totally different evolutionary line.

The last uintatherium finally drags itself out of the mud hole and slowly sets off to catch up with the rest of the herd, munching tufts of grass on the way. Leaving this scene behind, let's move halfway around the world to a beach in Pakistan.

A four-legged whale

A magnificent beach stretches before us for as far as the eye can see. This area, destined to become a desert, is now an arm of the ocean, with crystalline, azure waters. It's a clear, windy day.

A strange animal is plodding across the strand. It is about the size of a cow but has quite different limbs: they are much shorter and stockier, and faintly resemble flippers.

Before we can get a better look at it, the animal has slipped into the water. But it is obvious that it is a mammal. Approximately 300 million years after its distant vertebrate ancestors came onto dry land, this creature is doing exactly the opposite: it is returning to the water to live.

This is the *Pakicetus,* one of the forebears of the whale. Its name indicates that it was found in Pakistan (hence "Pakistan whale") and that it is considered of the line of the cetaceans.

It certainly looks more at home in the water: its movements are agile and it is obviously better suited to marine than to terrestrial conditions. For a moment we get a glimpse of a row of terrible teeth, like those of a killer whale. But then, after taking

Some huge animals wade into the lake. From afar, they look like rhinoceroses, but these uintatheria actually belong to a group whose origins are not yet clear.

a deep breath, the animal arches its back and dives into the deep to do some hunting.

Thus, the "squirrels" and "mice" of old managed to give rise to this huge beast as well. After generating a line of land ungulates (that is, animals with hooves, like present-day cows, goats, antelopes, and horses), they very slowly transformed into "un-

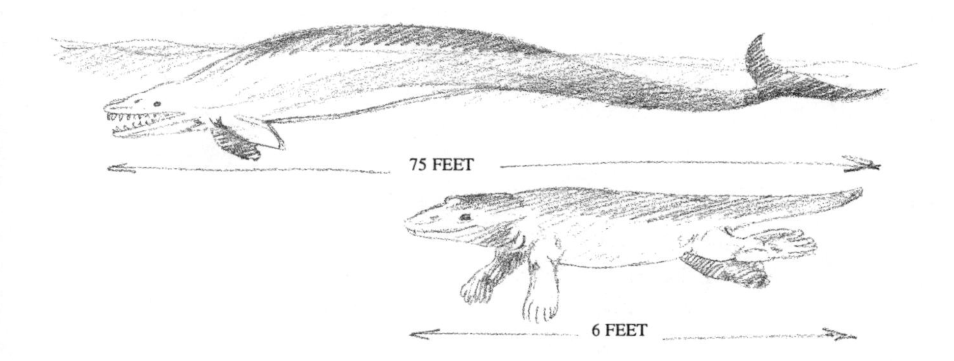

Top: The *Basilosaurus,* a surprising cetacean with the head of a predator (it is armed with long, sharp teeth) and the extremely long body (75 feet!) of a snake.
Bottom: A *Pakicetus,* one of the forerunners of the whale. Its name indicates that it was found in Pakistan and that it is believed to belong to the line of the cetaceans.

derwater cows," that is, into mammals (cetaceans) that started to live off the ocean's resources.

The *Pakicetus* (and other "divers" of the time belonging to the large group of *Archaeoceti* which includes approximately 140 fossil genera) gave rise to the largest animals existing on this planet, that is, whales. The blue whale, for example, measures ninety feet in length and weighs 130 tons.

Slowly, the descendants of the *Pakicetus* and similar animals increased in size and improved their proficiency underwater until they could go without surfacing for forty minutes and descend (like the sperm whale) to depths of over 3,000 feet, where the pressure is that of one hundred atmospheres.

In the course of 50 million years of evolution, this cetacean fauna will give rise to a variety of forms, including the *Basilosaurus,* which appeared in the oceans of North America about 40 million years ago. This was a very surprising cetacean indeed: imagine a cross between a killer whale and a snake, with a predator's head armed with razor-sharp teeth and an extremely long, hydrodynamic body. It reached lengths of up to seventy-five feet.

When the first fossil remains of the *Basilosaurus* were found in the last century, they were taken for those of a dinosaur. There was, in fact, a marine reptile that swam these oceans during the age of the dinosaurs: the mosasaur. Yet, although it strongly resembled the *Basilosaurus* and probably had the same snake-like way of swimming, it was just another example of evolutionary convergence.

A few minutes have gone by and the *Pakicetus* is coming up for air in the distance. Its autonomy underwater is still limited, but the cetacean line has begun and natural selection will see to shaping animals that perform increasingly well underwater. But natural selection is already shaping many other mammals, such as the greatest mammalian predator that ever existed, a terrifying animal whose story has yet to be entirely told. To get a better idea of what we are talking about, let's fly to the vast plains of Central Asia.

DECEMBER 28TH (38 MILLION YEARS AGO)

An eighteen-foot wolf

A howling wind is sweeping over the Mongolian steppe. Grasslands extend to all sides like a sea; our tent is the only visible craft. Huge clouds roll across the sky, a rainstorm is brewing.

We have been camping here for several days in hope of catching sight of what is considered the largest carnivorous mammal of all time: the *Andrewsarchus*.

Although it vaguely resembled a wolf, the *Andrewsarchus* was much larger: six feet at the shoulders with a three-foot-long head.

Finally, we spot one in the distance coming over a hill. And it's not alone; another one pops up behind it and then another. There are three in all.

At a lumbering trot, they head down the hill toward a small lake not too far away. Their gait is awkward, but it reveals a powerful body. Seen from afar, they look like a cross between a wolf and a hyena, but they are actually much bigger: they stand at least six feet at the shoulders and are fifteen to eighteen feet long, almost the size of a small truck.

The beasts become more clearly visible as they approach. Their heads are particularly striking: they are almost three feet long by themselves, with long narrow jaws that make them look out of proportion to the rest of the body. Large, pointed fangs, suited for making short work of any herbivore, stand out in the half-open, panting mouths.

The beasts stop. The first raises its snout and sniffs the wind. We wonder what the prey of such an animal could possibly be. There is always sense behind the workings of nature: it would never create such a giant if only to hunt small animals; that would be a waste. If the *Andrewsarchus* is so strong and powerful, its prey must correspond. They could be large herbivores that graze the plains in large herds.

The one that seems to be the leader of the pack continues to sniff the air. Now it lowers its head in an attempt to pick up a scent on the ground. It takes a few steps in our direction.

Suddenly, a crackling flash of lightening illuminates the landscape and a crash of thunder like a bomb exploding jolts us. Terrified, the *Andrewsarchus* dash away as the skies open and the rains pour down, soaking us to the bone. In only a few seconds, the sky has become as dark as pitch. Meanwhile, the three beasts canter along the crest of the hill before disappearing down the other side.

Little is known about the *Andrewsarchus*. Analyzing the animal's skeleton, many researchers have come to the conclusion that it was not very fast and, therefore, unfit for the kind of hunting that calls for sprinting, as lions or wolves do. Some of its characteristics seem to indicate that it ate carcasses.

The surprising thing is that the *Andrewsarchus* is in no way related to modern carnivores (such as wolves, lions, and hyenas); it is far more likely to be related to their prey, the ungulates (the group including deer, pigs, horses, buffalo, and elephants). In fact, the dominant animals among mammal predators have been very different at different times. Let's take a quick look at them.

An alternating sequence of mammal predators

With the disappearance of the dinosaurs, the continents remained devoid of large predators. As seen, their niches were initially occupied by birds such as the *Diatryma* and the *Phorusrhacus*. Yet, these were special niches, limited in time and geographic range.

Soon, a first generation of predators emerged: the creodonts (meaning animals that like to eat flesh). They produced a large variety of hunter "models" (as many as fifty genera), which generally resembled foxes and cats and played the part of modern mongooses, wolves, hyenas, bears, and lions, although they were quite different from an evolutionary point of view. They also produced giant forms with skulls twice as big as those of a tiger (26 inches).

Yet, after undisputed dominion in the beginning, the creodonts were quickly superseded by a new generation of predators with a more agile anatomy and a larger brain. Toward the end of the Eocene, 40 to 35 million years ago, the ancestors of dogs, foxes, mongooses, genets, and felines appeared. Later, in the early Miocene, hyenas came on the scene.

One group, however, outclassed all the others in strength

and prowess: the *Eusmilus,* or saber-toothed tiger. This extraordinary animal survived for an extremely long time. The interesting thing is that the various kinds of saber-toothed tigers that succeeded each other actually belonged to different evolutionary branches; some were even marsupials. But the predatory "model," characterized by two highly developed upper canines, was obviously so successful that many types of animals converged toward it.

For anyone looking at these skulls today, it seems almost impossible that an animal could have benefited from such an apparently cumbersome "anomaly." But knowing the use that these "sabers" were put to (which was quite different from that of felines today) is the key to understanding their success. Let's take a close look at the way the "saber-toothed tiger" attacked.

To do so, we'll have to return to the North American plains, on a late afternoon in the Oligocene, 30 million years ago.

DECEMBER 29TH (30 MILLION YEARS AGO)

The "saber-toothed tiger"

A light breeze has finally started to cool the air after a hot summer's day. Life is picking up close to the river, with long-legged wading birds strutting along the banks in search of prey to "harpoon" with their beaks.

A solitary male brontothere lumbers slowly down to the water alone. Cautiously its tiny eyes scan the surroundings. The Y-shaped horn on its nose catches on a low-hanging branch and finally breaks it, but the brontothere takes no notice and slowly trudges on.

The long, lethal white fangs of the "saber-toothed tiger" sink into the skin of the brontothere.

With a total body length of over twelve feet, the animal stands about seven and a half feet high. And yet, this giant knows that its size is not necessarily enough to defend it from the most ferocious of all predators: the formidable "saber-toothed tiger."

This place, with its trees and vegetation, is perfect for an ambush. The brontothere must have sensed something. It stops dead in its tracks and stands motionless for a long time before snorting and continuing on its way.

Suddenly, a shadow shoots out from the bushes. The brontothere barely has the time to turn its enormous head, before a huge mass of fur, fangs, and muscle flies toward it like a torpedo. The tiger, about six feet long, keeps its head low as it races toward it victim.

It is almost as though we were watching the scene in slow motion: we can see the muscles on the beast's shoulders and haunches rippling at every step, first contracting and swelling and then extending to project the powerful body forward.

The brontothere drops its head, waiting to meet the predator with lowered horns. But it is too late. The tiger has already avoided the mighty horns and with one powerful leap is upon its prey, its jaws gaping, its claws outstretched.

The impact is violent as the tiger lands somewhere between the brontothere's haunches and belly. Its front claws anchor into the flesh and its legs bend to soften the impact, while its two long, white "sabers" sink home. The jaws close, and with a violent downward shake of its head, the tiger rips a piece out of the victim's side. A deep, broad strip of bleeding, living flesh is laid bare.

The brontothere vainly tries to shake the animal off by whirling around and stabbing at it with its horns. But the tiger knowing it has already won, loosens its grasp and withdraws to the bushes.

Wheezing, the brontothere tries to charge, then stops, its horns lowered. It does not seem to have realized exactly how deep the wound is. First, it tries to limp away, then it loses its balance.

Its wheezing becomes louder and more frequent as the pool of scarlet blood collecting under its body spreads. How long will it be able to hold up? Not for long.

Two eyes from the bushes continue to glare at the floundering animal. Just below them, two "sabers" already dripping with blood are ready to attack again.

The judo wrestler and the samurai

A series of clues that we will look at in a moment suggests that this is how "saber-toothed tigers" fought. But first, we would like to clarify that the term "tiger" is incorrect (even though it is commonly used). These animals are not felines, like cats, pumas, lions, and leopards. They belong to groups that have disappeared.

Even the marsupials (like the creodonts mentioned previously) developed a form of "saber-toothed tiger": the *Thylacosmilus*. This "model" must have been very efficient, as it was adopted by very different groups of mammals.

Although originating from diverse evolutionary lines, the many "saber-toothed tigers" all developed the same characteristics: much more powerful front legs than hind legs (like the hyena); a short snout; an amazingly strong and muscular neck; a jaw that could open more than 90 degrees (much greater than the aperture of a lion, which reaches a maximum of 65 degrees when it yawns); and, above all, extraordinarily developed upper canines—two slightly curved and razor-sharp fangs.

What purpose did these adaptations serve? A comparison of the hunting techniques of one of these "tigers" with those of a lion will answer that question. A lion "anchors" onto the neck of its prey with its upper canines and snaps its lower jaw shut like a trap, strangling the prey or breaking its spine.

"Saber-tooth tigers" did exactly the opposite: the lower jaw anchored onto the victim's body, while the terrible "sabers," activated by powerful neck muscles, came down like a guillotine.

"Saber-toothed tigers" would tear away huge pieces of flesh (approximately eight to ten pounds at a time), leaving immense wounds that eventually led to death from hemorrhage, while the perpetrator waited on the sidelines. In short, the lion uses the tactics of a judo wrestler, who tackles his adversary and then strangles him, while the "saber-toothed tiger," like a samurai, brings down its victim with one fell blow.

But why does the samurai not tackle its adversary as well? Because its adversaries are different. While lions attack small and medium-sized animals (the largest being the buffalo) that can be immobilized and strangled, and do not bother with such "living fortresses" of the savannah as elephants, rhinoceroses, or hippopotami, "saber-toothed tigers" were specialized in killing the largest mammals, such as mastodons, mammoths, dinotheres, and brontotheres. These animals cannot be tackled; they can be brought down in only one way: by wounding them mortally in a vital spot such as the belly, throat, or haunches.

When these large (and slow) animals became extinct, "saber-toothed tigers"—too specialized in hunting this particular kind of prey—also disappeared.

Two thousand "tigers" in the tar sands

An interesting fact has emerged from the study of *Smilodon,* a "saber-toothed tiger" that lived in California ten thousand years ago. The remains of over two thousand *Smilodon* have been found in tar sands together with those of thousands of other animals. All were probably attacking some large prey that was already bogged down when they got trapped in the sands themselves. Study of the bones has shown that many of the "tigers" were maimed and therefore probably on the lookout for "easy" prey. Nevertheless, paleontologists have found that the maimings (fractures of the limbs, luxations, bone infections, arthritis, damage to the spine or aftereffects of bites from other "tigers") were often serious enough to call for the aid and solidarity of the other animals of the group, at least during the time immediately following the trauma. In other words, unlike modern tigers, "saber-toothed tigers" were not solitary animals; they lived in packs, as lions do today, and helped disabled members, probably by giving them access to prey killed by others.

DECEMBER 28TH TO 31ST (35–2 MILLION YEARS AGO)

The largest mammal

While the various versions of the "saber-toothed tiger" covered a very long period of time (almost 25 to 30 million years), other mammals had their ups and downs: they appeared and disappeared, were transformed, went extinct, changed.

In the approximately 30 million years that separate the Oligocene from modern times, the evolutionary branches of life created an incredible wealth of forms. It is in this period that animals with extremely different characteristics appeared, the animals that were to lead to elephants, camels, dolphins, deer, horses, bats, monkeys, moles, bears, anteaters, pigs, seals, giraffes, and others.

This period also gave rise to many strange forms that were to reach us only through the fossil record. The diversity of animals that we see in a nature park or zoo or even in television documentaries is nothing compared to the infinite range which nature's inexhaustible imagination produced in the course of millions of years.

In some ways, this variety is simply a continuation of the old Lego game: the assembly that started with molecules in the primordial broth, continued with cells in the Archaeozoic and with colonies of cells and tissues in the Cambrian, and then went into the mixing of genes and characteristics throughout evolution.

One of these "assemblies" led to an incredible individual that is now grazing not far from us: this is the largest mammal ever to walk the earth, a *Baluchitherium*. To see it, we have come all the way to this treed savannah lying between Pakistan and China. It's a rather hot, flat plain. But the vegetation is very dense here, as there is a small stream close by. Strange white birds with long tails soar overhead. A large wader sifts the water through its long bill in search of food.

The *Baluchitherium* is quietly and methodically munching the leaves from the top of a tree. To have some idea of its size, imagine a locomotive: this animal is twenty-four feet long and sixteen and a half feet at the withers (as high as a two-storey building). A horse-like neck stretches up from that point, topped by a head the size of a desk (which looks small in relation to the rest).

What does this animal resemble? It's hard to say. It could be related to any number of animals: it has the legs of an elephant, the body of a horse, the eyes of a rhinoceros, the partially hollow bones of a dinosaur. It weighs thirty tons, as much as four or five elephants together. With its prehensile lips, it grazes like a giraffe, reaching leaves that grow up to the incredible height of thirty feet (between the third and fourth story of a building).

This *Baluchitherium* is not alone; there are others behind it, also intent on the tender uppermost leaves of the trees.

A young male is suddenly driven from the herd by a female. She was probably fed up with his advances. But the smaller animal won't be put off: he trots up beside her and tries again to stop and mount her.

Opposite page: A *Baluchitherium* is more or less the size of a locomotive: twenty-four feet long and sixteen and a half feet at the shoulders

His "antics" finally draw the attention of the large *Baluchitherium* closest to us: turning from his grazing, he lets out a deep roar as he charges at his rival. We can feel the ground vibrating under his pounding hooves. The young male realizes that there is no time to waste and gallops off, the dominant male on his heels.

It will not come to a fight; the leader's challenge, followed by an act of submission by the subordinate, is all that group ritual demands to reestablish order. The two animals stop at a certain distance from one another. The matter has been settled. Both can return to their grazing.

Basically, these *Baluchitherium* have taken over the niches of the large herbivorous dinosaurs such as brachiosaurs and have grown to an equivalent size. But they, too, will disappear in time, leaving behind their bones to astound the paleontologist.

Yet, the *Baluchitherium,* a strange mix of rhinoceros, horse and elephant, is not the only oddity of this age of the great mammals. Many other bones belonging to animals that lived between 30 and 2 million years ago are still puzzling to paleontologists. Some seem to be very unusual and incomprehensible "assemblies"; two examples are the *Chalicotherium* and the *Macrauchenia.*

Strange "assemblies"

The *Chalicotherium,* which lived in the lower Miocene (approximately 20 million years ago) looked like a cross between a horse, a bear, and a gorilla. It had enormous claws, which it may have used to pull down branches so that it could eat the leaves.

The *Macrauchenia,* an inhabitant of South America during the Pleistocene (approximately 2 million years ago) resembled a camel, but had the paws of a rhinoceros and a small trunk.

Odd "assemblies"? Let's say imaginative. Nature never does anything without a reason: each organ, each feature, corresponds to a function, an adaptation. Nature's imagination is at the service of natural selection: it constantly proposes new and different solutions that are then rewarded or penalized by the environment.

Another example is the *Platybelodon,* an animal with a large shovel-like trunk that lived in the upper Miocene, between 11 and 6 million years ago. It was basically an elephant, but its

The *Platybelodon*. The animal sports a trunk equipped with a large shovel for digging up plants.

trunk was graced with a pair of flat teeth at the end that could be used as a huge spade to dig up grass and other plants inaccessible to other animals. That this tool was a great success is attested to by the fact that the *Platybelodon* spread throughout Europe, Asia, and Africa and existed for over 5 million years. But, as the history of evolution teaches, all species are both winners and losers because even the best-adapted still run the risk of environmental or climatic changes.

These were not the only changes, however, to which apparently powerful animals that dominated their environments were vulnerable. Even small geographic changes could at times create serious upheaval and lead to extinction.

This happened to the *Megatherium,* which lived in South America in the Pleistocene, between 2 million and six thousand years ago. It was another one of those giants that roamed the tree-covered savannah feeding on leaves. The *Megatherium* was eighteen feet long, weighed three tons, and had long, lethal claws. One blow could produce considerable damage. It was probably a very slow animal, a little like present-day sloths; but no other animal in its environment could eliminate it, much less cause its extinction.

The *Megatherium:* another leaf-eating giant (eighteen feet long) that roamed the tree-covered savannah.

Trouble began when a small geographic change took place: North and South America were linked by the isthmus of Panama. This brought about a sudden flow of fauna from north to south (and vice versa, of course). For the *Megatherium,* the arrival of competitors that had not been shaped by the local environment completely overturned the situation and triggered new selection.

Finally, the most agile predator of all—man—arrived to take over its niches. Modern man, that is, North American natives, who drifted into the region around eleven thousand years ago, saw the last *Megatherium* and probably hunted them: bits of fur from this giant have been found in a cave dating back six thousand years.

7:30 P.M., DECEMBER 31ST (2 MILLION YEARS AGO)

Galloping for over 50 million years

The end of our excursion among mammals brings us almost to the end of our journey. Human beings are making their entrance. The Pleistocene (which extends from about 2 million to ten thousand years ago) is the period in which humans will undergo extraordinary development. As a consequence, the next chapter will be dedicated entirely to the greatest revolution of all, the development of creative intelligence and, therefore, the advent of human beings.

But on this last evening, the 31st of December, we have come to the plains of North America to meet an old friend: the horse. The horse, which will have such importance in human history, was born on the North American plains 2 million years ago, at almost the same time as the birth of *Homo habilis* in Africa.

A herd of horses is grazing on the grassy hill land around us. The grass is an intense green, the sky a vivid blue dotted with small floating white clouds. There are some foals in the herd; one is being suckled by its mother, a strong, beautiful mare. These horses are identical to the ones we raise in our stables today, only smaller and stockier.

This is no chance encounter. We have come here because quite a bit is now known about the evolutionary history of horses.

It started about 50 million years ago with the *Eohippus* (the "dawn horse"). The *Eohippus* lived in the forests and, like all forest animals, was small. Although no larger than a dog, it already had a typically equine appearance (although it may have had a small trunk, a fact which is little known).

The progressive disappearance of the forests (for climatic reasons) and the emergence of the prairies called for new adaptations, such as the ability to run in open spaces to escape predators.

The horse evolved in this direction: study of its bone structure shows that over a period of 50 million years (*Epihippus, Miohippus, Parahippus, Merychippus, Dinohippus*), the bones of the leg were gradually transformed and the number of toes decreased. This evolution allowed the horse to adapt to running, making it one of the fastest animals on Earth and the one able to cover the longest distance in the shortest time.

The horse closest to us has stopped grazing. Raising its head, it tilts it slightly to one side to give us a curious look. This is obviously an intelligent animal, but it has no way of knowing that on another continent, thousands of miles away, another mammal is evolving that will one day climb on its back and make it jump hurdles in shows, race on tracks, march in parades, and charge into battle.

We take a few steps forward. The horse sidesteps. It is not afraid of human beings; it does not know them. But it is cautious. With a shake of its head, it trots off back to the herd.

The evolution of animal intelligence

According to a common interpretation of available fossil skulls, there has been an evolution in the brain of mammals in general and in the brain of the line that led to the horse. In fact, a relative increase in the size of the brain of mammals—and, therefore, in a broader sense, in their intelligence—seems to be observable, even after taking account of differences in the size of the many mammalian "models" that have succeeded each other over the last 50 million years.

Although this issue is rarely discussed (brain development is usually considered only in relation to human development), it is characteristic of the entire evolutionary process.

In other words, not only has there been a gradual increase in brain volume throughout evolution, but this increase in brain mass (and thus in the number of neurons capable of processing information) has also taken place *within* the individual species, albeit at different rates: almost zero in reptiles, very slowly in birds, slowly in mammals, quickly in man.

Therefore, a new kind of selective pressure is about to be exerted: selection for intelligence.

After having developed in all directions (speed, hearing, smell, vision), mammals have reached the point where they can now start to evolve in this new and important direction. The "critical mass" reached by neurons (and the concurrence of several favorable factors) is making it possible for the brain to use and process environmental data in an increasingly effective way. This development will, eventually, lead to the selection of intelligence and creativity.

The herd of horses has started to move: a small group sets off at a gallop, and the others soon follow. In a few minutes, all have disappeared behind a clump of trees on the hill, leaving the silent prairie reverberating to the low rumble of hooves.

Later, horses will become extinct on the North American continent. But others will already have crossed the tongues of land spanning the Bering Strait between North America and Asia, to spread from there into Europe.

Only many thousands of years later, in the sixteenth century, will their descendants return to their place of origin by ship, along with the conquistadors, that is, the descendants of *Homo sapiens sapiens* who are to make their appearance on the planet at about "ten minutes before midnight."

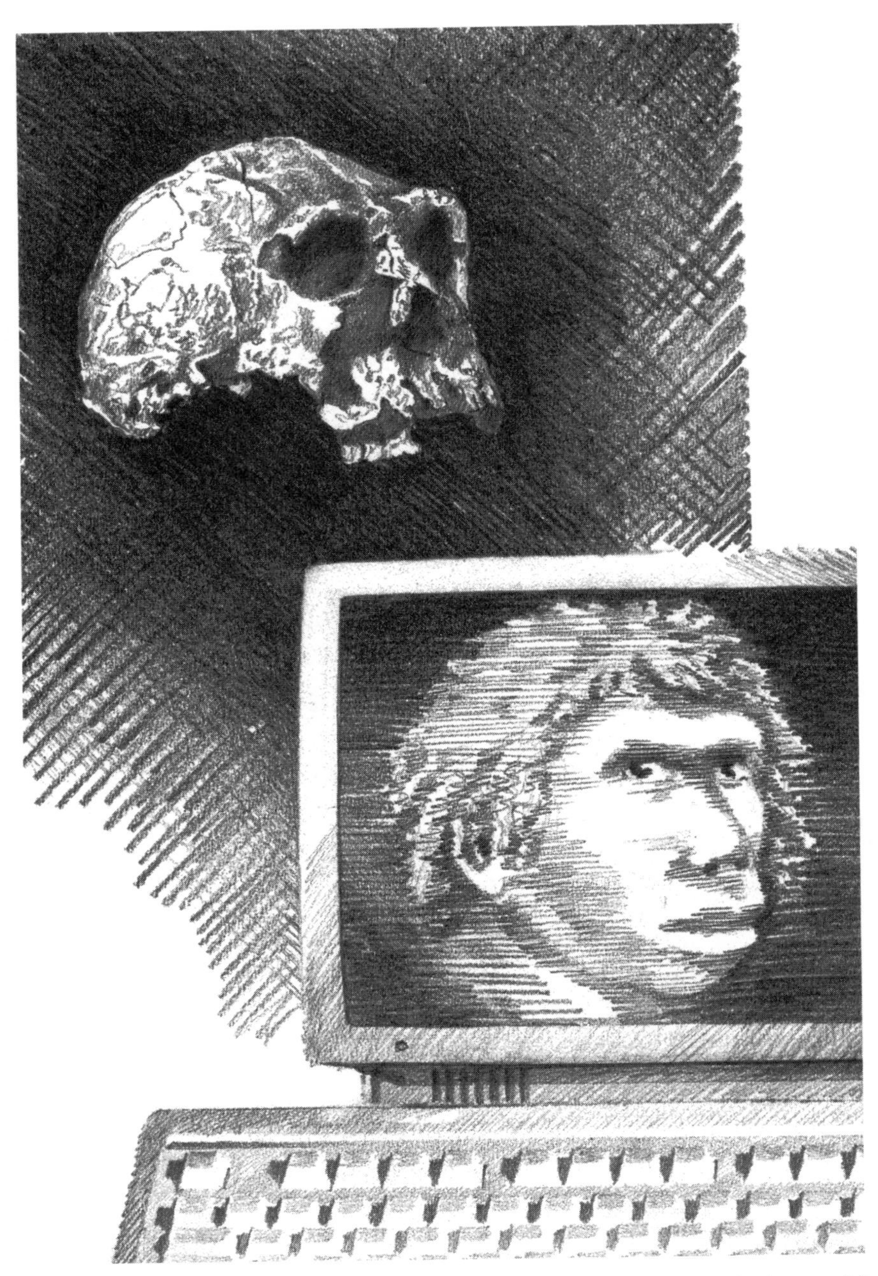

"Genetic research, dating techniques, comparative anatomy, evolutionary biology, and study of the fossil remains and tools available make it possible to sketch out the evolutionary path that led to modern human beings."

17

A Thinking Machine

11:50 P.M., DECEMBER 31ST (80,000 YEARS AGO)

A camp of sapiens sapiens

It is ten minutes to midnight on December 31st of our calendar when we reach the edge of a rocky plateau in the African savannah. These are the last ten minutes of evolution.

A column of smoke rises not far below us; it comes from a camp of *Homo sapiens sapiens.* A small group, no more than thirty individuals, is camping around simple thatched huts. We can see them quite distinctly through our binoculars: men and women, all rather young, and some children.

A slightly older man who looks like the leader is shouting at three younger men. Is he reproaching them or could this be his normal way of talking? Still shouting, he now shakes a long pointed stick—a spear. Their discussion probably has something to do with hunting.

In the meantime, two women are cooking pieces of meat over the fire. Close by, some men are carefully scraping the skin of an animal (perhaps a small gnu) with some sharp rock chips.

No one is wearing any clothes or furs: the climate is mild and even the nights are balmy on the savannah. The pelts will probably be used as pallets inside the huts.

This is a nomad camp: these people are hunter-gatherers who move with the flow of migrating animals. In this season, the savannah is lush with grass and hosts large herds of gnus and zebras which can be seen grazing in the distance.

A rudimentary container is piled high with the tubers gathered in the vicinity by the women while the men were out hunting.

The sky is a deep pink. The low cloud covering has been set ablaze by the sinking sun, silhouetting the elegant umbrella-shaped crowns of the acacias on the horizon.

From the camp comes a rhythmic tapping: a man sitting close to one of the huts is working some stone. The boy seated beside him observes silently as a tool emerges under his able blows. The method is already very sophisticated and produces a variety of tools: burins, gravers, punches, scrapers, handled blades, and "willow leaves," each serving a specific purpose.

This tiny tribe is "modern" in all respects. It knows how to make fire, hunt, and build shelters. Above all, it already uses evolved language to communicate, hand down learning, organize social life, and use ever more complex symbols and abstractions.

These *sapiens sapiens*, who originated in Africa at some unknown time 130,000 to 100,000 years before the modern age, pushed increasingly northward and, at a certain point, spread throughout the world.

Genetic research shows that the first *sapiens sapiens* started their extraordinary migration about 60,000 years ago. This expansion was to take them to Europe (40,000 to 35,000 years ago), Australia (40,000 to 30,000 years ago), Japan (30,000 years), Siberia (40,000 years), and the Americas (12,000 to 10,000 years).

In the course of their migrations, these originally dark-skinned human beings diversified. As they spread throughout the planet, the color of their skin, their hair, and their eyes changed. But they were and still are all *sapiens sapiens*: the same species, the same stock.

Their success was so staggering that it soon led to the disappearance of the various other evolutionary lines of *Homo* that had migrated to Europe and Asia in much earlier times—*Homo erectus* and their many ramifications (one of which developed into the Neanderthals in Europe).

Homo sapiens sapiens, who originated in Africa, is a new, much more evolved line. It is the terminal point of a slow process of transformation that led to the emergence of a life form which, unlike all others, started to understand the world in which it lived. How did this amazing story unfold?

Going back 4 billion years

The path that led to modern human beings is not easy to reconstruct in detail and certainly not in the space available here (we have attempted to do so in another book*). But many pieces of the puzzle are now beginning to fall into place and provide us with a rather convincing overall picture.

Despite the many gaps that still remain, genetic research, dating techniques, comparative anatomy, evolutionary biology, and study of fossil remains and tools make it possible to sketch

*Piero and Alberto Angela, *The Extraordinary Story of Human Origins* (Amherst, N.Y.: Prometheus Books, 1993).

out the evolutionary path that led to modern human beings. For the sake of brevity, only a few essential passages of this route will be mentioned.

Modern human beings have two certain ancestors: *Homo erectus* and before them (approximately 2 million years ago), *Homo habilis*, who were the first to manufacture stone tools.

Beyond them, it becomes increasingly difficult to establish who's who in the human genealogical tree. Fossil remains present us with a large number of possibilities, starting with the Australopithecines (3 to 4 million years ago), even though researchers still do not agree on the precise line of descent.

Beyond the Australopithecines, the line leads to a host of animals (dryopithecine) that branch out in various directions, to the *Propliopithecus* and the *Aegyptopithecus* (approximately 30 to 35 million years ago), a fox-like animal with a brain the size of a ping-pong ball.

Continuing back we come to the *Purgatorius* (approximately 70 million years ago), a kind of arboreal squirrel which many believe to be the mammalian forefather of the evolutionary line that led to primates and human beings.

The *Purgatorius* was one of the many small mammals that existed at the time of the dinosaurs and which we discussed in an earlier chapter; for almost 150 million years, they were "oppressed" and forced to live a "clandestine" life by the large reptiles that occupied all environmental niches.

Going back even farther (more than 230 million years), we come to the evolutionary lines (reptiles, amphibians, fish) that take us to the oceans of prehistoric times and, farther back yet, to the first pluricellular beings, to single cells and to the lipid vesicles in the primordial ponds. They are all our ancestors.

A very long and imaginative route. Who knows whether we will ever discover its many ramifications and details? One thing is certain: records of it are still contained in the chromosomes of each and every one of us.

Human chromosomes: a patchwork quilt

The DNA found in our cells descends from the DNA of primordial times. We have "collected" many of the genes that have proven successful in the course of natural selection and have incorporated them into our genetic "makeup." Most of our genes are very ancient (we spoke of this in relation to certain enzymes). In fact, they are almost identical—in the sequence of their units—to those of more primitive animals, such as insects. For instance, the genes of ribosomal RNA (the backbone of the organelle called ribosome involved in "reading" genetic messages) is almost identical in humans and flies. This is not surprising: we share with prokaryotes—both bacteria and archeobacteria—a host of DNA sequences (that is, individual genes) that code for proteins involved in the "housekeeping" of the cell. The implication is that these "shared" gene sequences existed in the common ancestor of all life forms prior to the early evolutionary splitting (or branching) that gave rise to the enormous variety of life forms that exists on the earth today.

Our chromosomes are a genetic "assembly," a patchwork quilt, composed of genes picked up at various times and on various occasions. Some have been preserved almost as they were originally, while others have changed at different rates. It is these genes, which undergo continual change, that create the variety of forms and "attempts" which are then subject to natural selection.

An enormous effort is being made to "map" the genetic material of human beings (the "genome project" uses powerful electronic equipment to automatically sequence human DNA). At the same time, another line of research is trying to assess the "average rate" at which mutations occur in nature (and in human beings).

When these studies will have been concluded, it will be easier to reconstruct human history and our kinship with other living forms. Not only that, by extending these studies to other organisms, we may be able to compare any two living species and calculate the evolutionary distance between them, that is, the distance of both from their closest common ancestor. Put another way, it will be possible to assess at what stage the two species began to diverge from a common ancestor. A large number of these comparisons may enable the reconstruction of the earliest branching event (dating back about 3.5 billion years), thereby establishing the root of the "universal evolutionary tree."

Of course, it doesn't take much to realize that a cat is closer to a tiger than to a dog (but also that it is closer to a dog than to a chicken and closer to a chicken than to a jellyfish). But researchers hope that if these differences become evident at the genetic level, it will be easier to understand the many branchings of evolution and to construct a global evolutionary (or phylogenetic) tree.

A recapitulation of life

The genetic patchwork contained in our chromosomes can also be observed in another way, during development of the fetus. In fact, the fetus (or the embryo) provides a showcase for the "common heritage" of such very different creatures as humans, dogs, chickens, and turtles.

In a series of observations carried out on diverse embryos in the nineteenth century, German biologist Ernst Heinrich Haeckel discovered a surprising similarity in embryonic development. The embryos of the mammal, the bird, and the reptile used in his experiment all started out with a fertilized egg that divided itself and gave rise to a morula (that is, a mass of cells that looks like a blackberry). And all then followed an astoundingly similar path. Haeckel called these various stages in the development of each new individual a "recapitulation" of the history of life on Earth.

In the first stage, in fact, each egg lives without oxygen, as was the case in the primordial seas (even the fertilized human egg is initially without oxygen, until it descends the tubes and implants itself in the uterus and starts to receive oxygen from the mother's circulation). Later, embryonic development goes through various other stages that "recapitulate" the various stages of evolution: fish (very rough gills even appear in the human embryo), reptile, ancient mammals, and so on.

In some way, the stages of evolution (phylogenesis) reappear during the construction of each individual (ontogenesis). As the well known formula states: ontogeny recapitulates phylogeny.

A brain with archeological layers

The most interesting thing that Haeckel noted was that the human brain also retraces this evolutionary itinerary. In fact, the successive layers of the human brain reproduce the archeology of the nervous system. Or, as some researchers have put it—using a metaphor that has been very successful with the media—coexisting with our typically human neocortex is the brain of a reptile and the brain of a very ancient mammal.

Haeckel's drawing, reproduced on the next page, shows the resemblance during embryonic development between the brains of a turtle, a chicken, a dog, and a human being.

Of course, this does not mean that evolution "superimposes" new things on existing structures (as when we put on underwear first, then a blouse, a sweater, a jacket, and so on while dressing). As we have seen, do-it-yourself nature tends to alter old parts to

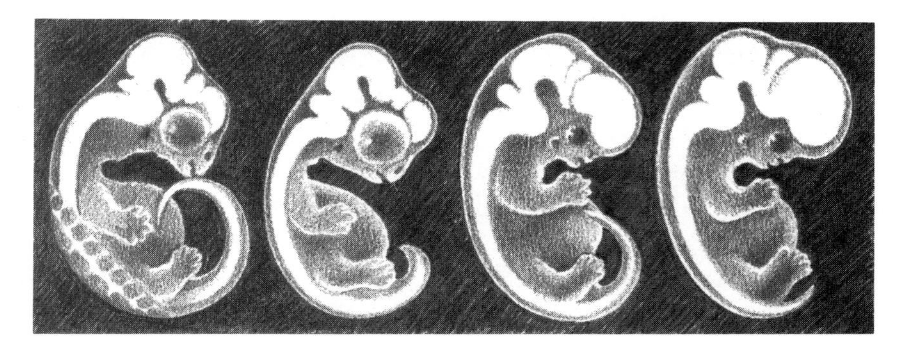

The surprising similarity during embryonic development between the brains of a turtle (sixth week), a chicken (eighth day), a dog (sixth week), and a human being (eighth week). (Haeckel)

serve new purposes, changing the "significance" of certain structures. This is similar to what happens in language: certain words take on different meanings depending on the context.

An exceedingly "mechanical" interpretation of Haeckel's observations has been criticized for this reason. Nevertheless, brain anatomy provides a striking illustration of the basic resemblance between the brain of a human being and that of a reptile, a bird, or a mammal, and this is evident during embryonic development.

Comparison also shows that the differences in the human embryo translate above all into an increase in that part of the brain destined to become the neocortex, that is, the part of the brain that processes, classifies, and associates information into a kaleidoscope of mental "assemblies."

The illustration on the following page very schematically shows how the various parts of the brain differentiate once development is complete. The neocortex (in black) is already present in the reptile and rather developed in the primitive marsupial. In human beings, however, its development is much more pronounced and requires convolution to fit into the available space (the same trick is used in radiators).

The development of the neocortex (in black) in a reptile (left), a primitive marsupial mammal (center) and a human being (right). The reduction in the size of the human semicircular area dedicated to olfaction at the bottom is evident. (Romer)

It is also interesting to see how other areas have changed. In human beings, the olfactory centers (the oldest region of the cortex) are much smaller, while the other primitive parts have changed shape slightly but are substantially still the same.

An increase in brain volume

There is, then, considerable continuity in the evolution of the brain, continuity that is even more marked in the evolution from the first hominids to modern man. No anatomical comparisons are possible in this case because the brains of ancient hominids are not available, but we do have a lot of skulls from which we can glean important data, mainly concerning brain volume.

By measuring various skulls covering a period of two and a half million years, paleontologists have found a progressive increase in brain volume: from 450 cc in Australopithecines, to 600 to 700 cc in *Homo habilis,* to 800 to 900 cc in the first *Homo erectus,* 1000 to 1100 cc in the last *Homo erectus,* 1200 to 1300 cc in *Homo sapiens,* and 1400 to 1500 cc in modern human beings.

As we all know, brain volume is not everything: many other factors come into play. The size of an individual, sex, and even personal factors can produce remarkable differences in brain volume without affecting intelligence (the cases of famous authors with very different brain sizes, from 1000 cc to 2000 cc, are often quoted. These are very special cases, however, that have not been thoroughly investigated for the influence of pathological factors).

Important factors, in addition to volume, are neuronal architecture, that is, the distribution of nerve cells on the cerebral map; the way in which they are linked (or cabled) and their biochemistry; the number of synapses (the points of contact between neurons); and their "quality," since each synapse changes chemically with experience, especially during development.

The fetus of the human being and of the chimpanzee

But another very interesting observation has also been made in comparing embryos and fetuses before birth. A few years ago, evolutionary paleontologist Steven J. Gould pointed out the amazing similarity between the skulls of a chimpanzee fetus and of a human fetus. The difference, he noted, is that the chimpanzee's skull changes considerably after birth, while that of the human being maintains a shape much like that of the fetus (see the diagram on the following page).

This observation has triggered various considerations and hypotheses: in particular in human beings, the fetal stage seems

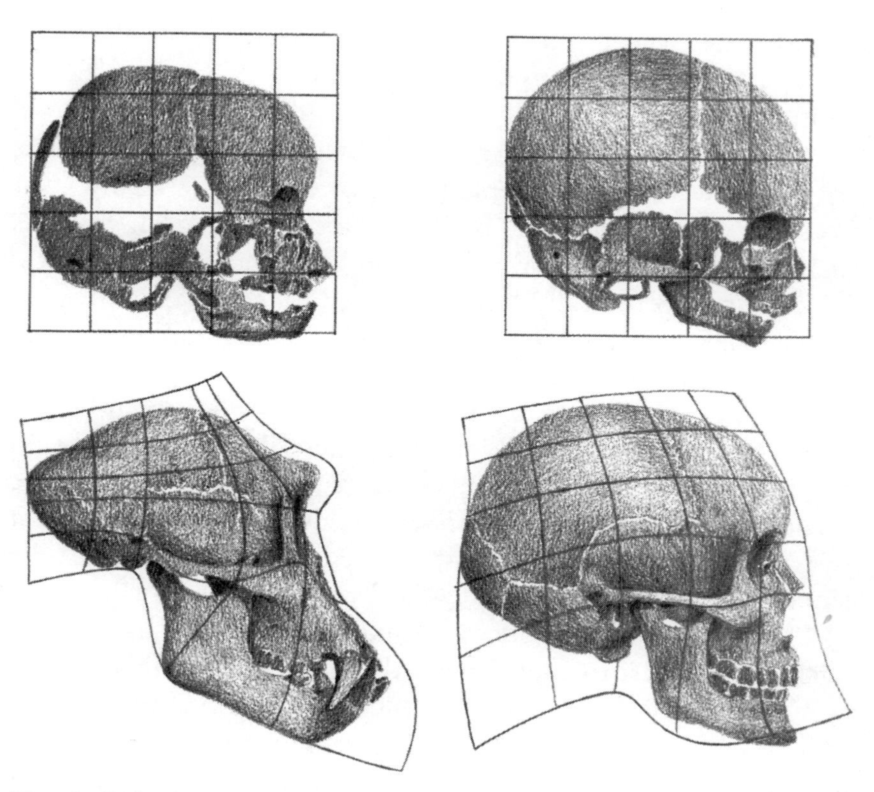

The similarity between the fetus of a chimpanzee and that of a human being can be seen in the two skulls shown at the top. As the chimpanzee grows, however, the shape of its face changes considerably while that of the human being remains basically unaltered (bottom). (Starck and Kummer)

to be "prolonged" beyond birth, offering a considerable advantage for the maturation of the brain and, therefore, the accumulation of culture. This prolonged exposure to environmental factors is important for selecting circuits, activating useful synapses to the detriment of the useless ones.

It is commonly accepted that a long infancy is a major advantage for human beings. But what does this mean from a genetic point of view? Is this prolonged brain maturation caused by certain genes that regulate the rate of development? Was that one of the key mutations that allowed for the emergence of a "cultural" species?

Study of the human genome, which we discussed earlier, may provide a better understanding of this in the future. But modern genetics is already making some surprising discoveries regarding similarities between human beings and chimpanzees. For example, genetically human beings are almost identical to chimpanzees: they are 99 percent the same! That makes us closer to chimps than horses are to zebras!

In the course of evolution, chimpanzees and human beings have accumulated extremely similar genes. Their genes were, in fact, identical up to 5 to 7 million years ago (approximately the time at which their common ancestor existed). Then, in the last part of their evolution, each introduced the differences into its genetic makeup that constitute the 1 percent difference.

What are these differences? Or rather, what have human beings introduced into their brains that is so extraordinary as to have completely changed their evolutionary direction?

The time may have come to penetrate directly into the brain of *Homo sapiens sapiens* in an attempt to understand the secret behind this great leap forward.

Exploring the brain of a sapiens sapiens

Now that the camp is asleep, we stalk quietly up to the huts. Some members of the group are lying near the fire, others are sleeping inside. A young man and a young woman are cuddled close together. A man with a thick black beard is snoring regularly. Beside him a woman lies sleeping, her arm resting on the baby curled on her stomach.

The bearded man looks like a suitable target. In fact, he even has a small scratch on the nape of his neck. It provides access to the circulatory system, which we are going to use to enter his body.

Once we have scaled down to cellular size, this cut looks like a gaping cavern. We walk into the cave just as the far end is beginning to be blocked off: dead cells and red blood cells mixed with fibrin are building a barrier that will eventually close the passage. A large number of white blood cells are also on the spot, destroying the bacteria and viruses that are trying to get in.

Pushing our way through, we manage to enter a tiny blood vessel. We are immediately picked up by the flow in the narrow tunnel. Red blood cells roll by in a flooding torrent. White blood cells and platelets bob up and down around us.

Here and there, strange cellular "monsters" hang from the walls like amoebas, continually changing shape to attack and absorb foreign bodies. These are white blood cells that are being activated into macrophages so that they can hurry off and fight infection.

The tiny tunnel finally empties into a larger one in which we can hear the heartbeat more distinctly: a dull, heavy thudding. Each beat causes a slight contraction in the wall affecting the flow.

We let ourselves be carried along by the scarlet torrent for a few endless moments. Each time the size of the blood vessel changes, the rate of flow changes. After being bumped and pushed by intense pulsations through the heart and lungs we finally start the long ascent toward the brain. Driven through a series of canals that branch off again and again, we finally enter the circulatory system of the brain. As the ramifications become more numerous, the flow carries us into a small tunnel in which we move at an appreciably slower pace.

Even this tunnel branches out into innumerable extremely narrow passages. The flow becomes even slower: we are in the brain, near one of the convolutions of the cortex in the frontal lobe area.

In front of us, a macrophage is trying to force its way into the brain through the wall of the blood vessel (probably to remove the "corpse" of some dead neuron). Given its ability to change shape, it finally manages to squeeze through a crack between two cells, like sand flowing in an hourglass.

This macrophage has come here through the body; in fact, macrophages use the circulatory system like a subway to carry them to the stations they have to defend from attack: the lungs, the spleen, the brain, and so on.

The "monster" has managed to squeeze almost entirely through the crack. Only a tiny appendage is still sticking out and even that is gradually being drawn to the other side. We decide to take advantage of the opening made by the macrophage to enter the brain ourselves, but it is already closing. We will have to hurry; not too much, though, for we, too, are foreign bodies, and the macrophage could decide to alters its destination and engulf us in its lethal embrace.

We squeeze our way through the cell membranes, which are fortunately rather soft and elastic like rubber. Beyond them are a number of "glia" cells, the last barriers to our passage. Finally, the greatest spectacle that nature has to offer—the human brain—opens out before us.

The cerebral rain forest

It's like stepping into a dense jungle, full of boughs, branches, roots, and vines. The "vegetation" is so dense that it forms one inextricable tangle.

It is not easy to identify the various neuronal "trees" and their branches in this cerebral rain forest. The terminal branchings of the pyramidal cells can be seen in this external part of the brain. These are cells with a rather long "trunk," which sinks into the cerebral cortex and has a tree-like crown of nerve endings (dendrites) at both the top and the bottom, just like a tree with its branches and its roots. Many boughs also branch out from the trunk and each one is covered with " thorns," that is, minuscule protrusions that look like buds. These are the points of contact with other neurons.

There may be as many as twenty thousand or more of these buds in a single pyramidal cell, like an immense switchboard. This gives an idea of the enormous possibilities for linkage among neurons. Plunging through the jungle's maze, we try to get deeper into the cerebral cortex. We can feel strange vibrations when we touch the branches: the electrical impulses running through the membrane.

It is now known that neurons use electrochemical signals to communicate. Traveling through the neuron, the electrical impulses reach the points of contact with other cells (synapse) and set off a chemical "spray" which in turn activates another electrical impulse. This is how the signal is relayed. Various kinds of chemical substances (so-called neurotransmitters) activate the synapses: acetylcholine, noradrenaline, serotonin, an so on. The combination of signals (which activate and inhibit), receptors, transmitters, chemicals, and cellular architectures gives rise to nervous activity.

Climbing down through the branches of the tree, we realize that there is another pyramidal cell below this one and another below that one, and so on. And there is another one beside it, too. They are all arranged in columns and layers: the apparent disorder actually hides an amazing order and structure.

Penetrating further into the cerebral cortex, we now see another smaller and more ramified type of neuron with a shorter "trunk." It looks more like a shrub than a tree. This is a stellate cell. It links the pyramidal cells and sets up a dense network of interconnections with the nerve bundles coming from the outside (there are six kinds of stellate cells of various shapes in different parts of the brain).

Descending the nerve shrubs we can see more pyramidal cells below us. They are also organized into layered columns. The arrangement of neurons resembles a woodpile made up of bundles of twigs standing on end, and with different layers formed by different kinds of twigs, all very ramified and interconnected into an endless, intricately linked maze.

But in this jungle of neurons, ramifications, and synapses, what are the differences that distinguish human beings from other living creatures? What allows this brain to think, imagine, and devise mathematics, music, and ideas?

Similarities and differences

Mice and other mammals have long been used for brain experiments. The brains of these animals have been studied and their cortex compared to that of humans. The results are rather surprising.

Let's start with neurons. The neurons of the cortex (pyramidal and stellate) are almost identical in human beings and in mice or other superior mammals. They are hard to distinguish under the microscope. In other words, the neurons of human beings are not special in any way; they are exactly the same as the neurons in the cortex of other mammals.

And that's not all: the density of the neurons is also the same (that means that there is the same number of neurons in one cubic millimeter of cortex).

What about synapses? Even more similar. The mechanism proved so efficient in the course of evolution that there was no need to change it in human beings.

But is there at least a difference in the density of synapses? No, it is the same per cubic millimeter. In other words, the "assembly kit" is almost the same for all. So where are the differences?

There are differences, but they seem to lie not in the *quality*, but in the *quantity* of the neurons. The human cortex is larger, thicker, and has more intercortical connections. It is also more developed in certain areas.

Comparison of the human cortex to that of other primates has shown that it is almost three times as large as the cortex of a chimpanzee and ten to twenty times that of other less evolved apes (and 150 times as large as that of a primitive insect-eating mammal). The greater thickness of the cortex in humans (especially in the area of sight) increases the number of stellate neurons, that is, those neurons that take care of internal linkages and that are, therefore, particularly useful for processing and associating information.

Increasing the size of the cortex increases the number of neurons. It has been calculated that the mouse has 65 million neurons, the chimpanzee 7 billion, and human beings tens of billions.

Finally and most important, in humans, this development has taken place in "strategic" areas, those set out for superior functions (such as language).

No one has ever counted the number of neurons in the human brain, just as no one has ever counted how many trees there are in the Amazon rain forest. But curiously, it seems that these two numbers are not far apart: both seem to be in the tens of billions. Considering that there can be thousands of synapses in each neuron (i.e., thousands of points of contact), the number of links possible in the cerebral cortex is enormous: between one hundred thousand billion and a million billion.

This network is so immense that it is practically impossible to calculate the number of possible pathways and combinations. It would be like having an alphabet made up of a million billion letters rather than just twenty-four. The number of new combinations is beyond our imagination.

Apparently, it is this "quantitative" increase in the number of neurons that has given humans the extraordinary "qualitative" advantage that distinguishes them today. It also underlies our ability to process incoming data.

Much remains to be learned about the structure and the functioning of this marvelous thinking machine. Research has already shed light on some areas, giving us an idea of certain basic principles. But one cannot help but wonder why some parts of the cortex developed in this way in human beings.

In attempting a reply, we could pose a number of related questions: Why is the area of the cortex (or rather, the paleocortex) involved in olfaction so surprisingly developed in certain animals? Or why is the part of the cortex involved in hearing so highly developed in others?

Just think of the dolphin. It has an extraordinary brain (even superior to that of humans in terms of the cortical size), but although also a very intelligent mammal, the dolphin has "specialized" in processing sonar signals.

Human beings, on the other hand, have specialized in processing memorized experiences. The most developed parts of the human brain are the associative areas, those of intercortical connections, and those for the elaboration of language.

In some way one might say that we are "underdeveloped" with respect to certain animals that have increased the number of neurons and brain circuits for memorizing and processing odors or sounds. But they are underdeveloped in comparison with us in other (far more important) brain activities.

The mechanism behind these differences in development is always the same. It is the mechanism linked to the crossed effects of heredity and environment that influences the evolution of all living beings in nature.

We have already spoken of this elsewhere in the book and will not go into it again here. It is enough to recall that heredity (i.e., genes) continuously offers the environment a variety of "models" of which the most suitable survive through selection. In this process, "successful" characteristics are accentuated by means of what is known as "selective pressure," which gives priority to the particular individuals (among the descendants) in which the successful characteristic is most developed.

The history of evolution is full of such "anomalous growths."

Looking at the animals in nature, it is easy to see that many organs (or parts of the body) have grown out of proportion: aside from the classic examples of the giraffe's neck and the elephant's lip-nose muscles, it is enough to think of the narwhal's tooth, the chameleon's and anteater's tongue, the gibbon's arms, the barn owl's eyes, the walrus's teeth, the ruminants' stomach, and the whale's lungs.

In comparison, the human cortex—as developed as it may be—is only two millimeters thick! It has not really grown that much, but it has grown in strategic places.

The tunnel leading to Beethoven

This is the drift of our thoughts as we leave the brain of the *sapiens sapiens.* To get out more quickly, we descended the acoustic nerve to the inner ear, crossed the eardrum, and are now walking down the auditory canal.

A glimmer of light is visible at the end of the tunnel. Only a few more obstacles to cross—some huge "plates" of dry skin and a couple of enormous hairs (which look like huge curved tree trunks)—and we will finally reach the outer ear and the outside world.

Our *sapiens sapiens* is still snoring; we can feel the vibrations rising through his Eustachian tubes. But as we trudge on, we are suddenly hit by a sound wave: Beethoven's Fifth Symphony.

Who could possibly be playing Beethoven's Fifth in the *sapiens'* camp? Curiosity drives us forward.

The music fades away and is gradually supplanted by a voice that sounds slightly metallic, as though it were coming through the loudspeaker of a radio:

"Good morning! Time to wake up. It's 7:30 A.M. You are en route. All onboard systems are okay. Happy Birthday from the captain's wife and daughter; they've put the birthday cake into the freezer until his return. Next call at 8:15. Over and out."

We reach the outer ear to find that we are on board a spaceship! The ear is still that of the *sapiens sapiens* in the camp, but the man with the beard who snored peacefully beside the woman and child is now stretched out in a swivel-back chair: he's the captain of the spaceship.

But . . . what time is it? It's midnight on our watches, midnight of December 31st. The last eighty thousand years have passed in only ten minutes. And we've come to the end of our journey.

12:00 P.M., December 31st

Through the porthole, we can see our earth shimmering in the distance: a tiny blue luminescent pearl with its white cloudy vortices floating in the darkness of space.

Visible below the clouds is the apex of a landmass, but we can't quite make out which continent it is. This is quite a different planet from the one we parachuted down to from our meteorite 4 billion years ago. This one is full of color, oxygen, trees, animals, and people.

We can also see the moon almost directly behind it: a white sphere suspended in cosmic shadow. Two billiard balls roaming through space. It is almost frightening to think that life needs only the slightest protective veil to exist and be able to wander tranquilly through the galaxy's icy solitude.

Right now, down there, under those clouds, people are playing cards, working, arguing, making love, investing in the stock market, sunbathing, dying, eating Italian food, campaigning, writing postcards.

Everything goes on as usual . . . as if the universe did not exist, as if the only universe were the one enveloped by that thin and fragile atmosphere protecting our planet and allowing life to breathe.

Many things have happened on Earth in the last ten minutes of our evolutionary calendar:

11:54 — *sapiens sapiens* started to migrate in all directions from Africa

11:55 — they arrived in Europe, wiping out the Neanderthals. At the same time, they spread to Siberia, Japan and Australia

11:57 — a *sapiens sapiens* artist left us the first (extremely beautiful) bone sculpture

11:58 — another great artist painted the roof of the cave at Lascaux

11:59 — a group of *sapiens sapiens* in Mesopotamia invented agriculture

11:59 30″ — the first Egyptian monuments appeared

11:59 46″ — Augustus was proclaimed Emperor (after less than four seconds, the Roman Empire fell)

11:59 52″ — the First Crusades were fought

11:59 56″ — Christopher Columbus discovered North America (and the *sapiens sapiens* populations that arrived there ten thousand years before him)

11:59 58″ — the French Revolution broke out

11:59 59″ — the first steam engine was brought into operation

11:59 59.5″—World War I broke out

11:59 59.8″—*sapiens sapiens* set foot on the moon

And where is *sapiens sapiens* headed now?

"It should not be forgotten that there are 100 billion suns in our galaxy (and that there are billions of galaxies in the cosmos)."

18

The Next 4 Billion Years

STENDHAL AND VICTOR HUGO

The cultural acceleration of the last minutes and especially seconds has been very impressive. Seen in evolutionary terms, these few seconds represent a giant step in the planet's history.

The possibilities lying ahead are so vast and diverse that no one can predict where current development will lead. The best we can do is to make inferences based on our observations and the knowledge available today.

In this concluding chapter, we will envision different futures in different time frames: a short-term future, a medium-term future, and a very long-term future.

The spaceship porthole provides us with a panorama of space and time: on one side is the earth, which represents our present; on the other is the universe from which we came and which represents both our remote past and remote future.

In fact, the earth is only a tiny roving speck whose evolution is linked to the future evolution of the universe as a whole. But let's start with the earth as it is today.

Someone once said that there are two ways of looking at the battle of Waterloo: one is Stendhal's way, describing the few square meters in which the men fought, swore, and died—a few dusty meters drenched in blood, sweat, and tears. The other way is Victor Hugo's way. His descriptions of the troop movements, the soldiers' assaults, and the cavalry's charges give a more distant and detached view of the battle, in which the individuals and their personal tragedies vanish.

These are indeed two ways of considering the same event, but these two ways complement each other, providing information at different levels. From up here in the spaceship, we cannot hear the cries of the soldiers, nor see their blood or tears. We can only perceive the troop movements, that is, the great trends and general direction of certain developments.

THE CULTURAL GAP

One trend seems to be dominating all others: the difference in the growth rate of culture and technology. This is probably the most destabilizing factor in human society in the short term. In an exceedingly short time, research has produced an incredible arsenal of technologies that are spreading throughout the planet, but this spread is not paralleled by the cultural growth needed to manage them properly.

Pollution, resource crises, overpopulation, and nuclear risks are only some of the main spinoffs of this continually widening "gap" between technology and culture. We can put it another way: generally, we do not deserve the technology we dessimate. But why?

An industrial (or postindustrial) society is not composed

solely of machines, fertilizers, power plants, antibiotics, electronics, bulldozers, and so on. It must also (above all) have widespread education; the ability to control and manage; suitable policies; various kinds of "antibodies" (ecological, ethical, behavioral, legal); "centers of excellence"; an understanding of the consequences of actions; the ability to self-correct; and rapidity in rebalancing distortions or possibly changing route. Without all these things (and many others), even the most useful technologies can quickly prove counterproductive and turn into instruments with which to destroy the entire planet.

This is the risk that human evolution faces in the short term. Suffice it to think of the population explosion that has been made possible by the spread of technology in the last decades. At the moment, the earth's population is increasing by 90 million inhabitants per year (as if more than the entire population of the United States were added to the planet every three years).

It is disconcerting to think that human beings have established safety limits for all their activities (only so many people are allowed into an elevator, an automobile, or a theater), but not for the planet they live on. Paradoxically, if the population continues to increase at the current rate for the next 500 years (only a few seconds on our calendar), the earth will be forced to host one person per square meter (including in the Sahara and Gobi deserts, the Amazon rain forest, Antarctica, the Himalayan peaks and the Siberian steppes, and other such inhospitable places).

Demographic projections estimate that world population will stabilize automatically between 10 to 12 billion inhabitants in fifty to sixty years. This means that young people today will witness a doubling in the population of the planet during their lifetime. That is an increase equal to twenty-five times the current population of the United States. If these new inhabitants (along with the old ones) wish to achieve a standard of living equivalent to that of the Western world today (even if only its poorer regions), the number of homes, jobs, hospitals, streets, schools, resources, and the quantity of food, transportation, industry, and energy required will throw the entire terrestrial ecosystem into crisis—especially in the absence of an overall context capable of controlling and managing that development.

In fact, each inhabitant brings with him/her an enormous burden of direct and indirect consumption (just think of the energy, resources, raw materials, and so on that a student consumes for twenty years without giving anything in return). Therefore, it is not solely a matter of space, as in a theater; it is a matter of equilibrium.

What kind of world?

We are not leaving our children a very fine legacy: we are handing down a world that is overpopulated and full of technologies that can be extraordinary if used wisely, but terrifyingly destructive if employed in the wrong way. The basic question is: Does the world toward which human societies are tending have widespread education; capabilities for control and management; ecological, ethical, behavioral, and legal "antibodies"; "centers of excellence"; the ability for self-correction; rapidity in redressing imbalances, and so on? Is it not instead a world increasingly equipped with more and more powerful machines, but basically archaic, pretechnological and prescientific, and full of tensions, conflicts, armed fanatics, poverty, and pollution to boot?

From the cockpit of our spaceship, the problem of the gap between the production and use of technology looks like a difficult one to solve. It is almost impossible to stop the development and spread of technology. Our only real chance is to educate human beings, using all available instruments in the most creative way. This prospect, however, is countered by inertia and indifference in both rich and poor countries.

Looming over everything is, of course, the specter of nuclear war (and nuclear terrorism). What kind of a future can we expect, then, in the next fifty to one hundred years?

VARIOUS FUTURES

The optimistic hypothesis that we favor is that human systems will slowly adjust and adapt to this technological "trauma" (albeit with periods of difficulty). Hopefully, new "targeted" technologies (in energy, production, ecology, and education) will be able to soften the transition.

Another hypothesis is that adjustment will be reached in a more dramatic way, after serious environmental crises, conflicts, major shortages, tensions, and isolationism.

There is also a pessimistic hypothesis, however. It involves nuclear holocaust or even the extinction of the human species. It is unlikely that this hypothesis will materialize. Human beings are worse (or better?) than bacteria: they cannot be eradicated from this planet, even in the event of nuclear catastrophe (which, fortunately, seems to be less threatening now than in the past).

But even if such an extinction were to take place, that is, if all the human beings (like the dinosaurs) were to disappear from the face of the earth, it would not be that dramatic for the planet. The earth did well enough without human beings for an extremely long time (from January 1st on our calendar to the evening of December 31st) and could easily do so again.

For billions of years there were magnificent sunrises and sunsets unwitnessed by man. The moon illuminated streams and waterfalls, animals played and mated, trees grew, flowers sent their perfumes out to the four winds. Nature will continue to do so without us. No one will notice our disappearance. And, of course, it cannot be ruled out that biological evolution will not gradually give rise to other intelligent species able to take our place. (A paleontologist once said that the most likely candidate for founding a new line in that case would be the raccoon, a very intelligent animal with prehensile paws and frontal vision. The predicted evolutionary time: four or five days on our calendar.)

Nevertheless, it seems very unlikely that the human species will ever be completely wiped out, in spite of all the crises it may face. After a few centuries of tensions and disequilibria, things should improve. What will the future be like, then, if human beings remain on the scene?

THE FUTURE: HUNDREDS, THOUSANDS, OR MILLIONS OF YEARS?

When we speak of the future, we generally think of the next century or, at the most, the next millennium, that is, times that are ridiculously short if considered on the evolutionary scale.

In fact, *a single day* on our calendar equals over *10 million years* on the planet. In other words, if the human species were to live "one week" beyond December 31st, it would have a future of over 75 million years!

In addition, that time would be amplified by cultural acceleration. It has been calculated that the earth's total "body of knowledge" doubles every ten or twenty years (more things have been discovered in the last one hundred years than in the preceding ten thousand). Even if this rate were to slow down somewhat from its present dizzying pace, millennia would still have the effect of millions of years, and millions of years the effect of billions. A single day on our calendar would probably be worth many millions of years in terms of cultural evolution.

The mind boggles when considering this kind of time frame. Not only are we unable to predict long-term developments, we are incapable of even vaguely understanding in what direction we are headed.

Today, we can see only some of the paths that will certainly open up and enjoy extraordinary development: space, of course; genetic engineering; and systems of artificial intelligence (with new and surprising applications: just think of the so-called virtual realities, which submerge us in realities that are artificial, illusory, but absolutely "real" for our senses).

Other "revolutions" are predictable: for example, a food revolution (we still eat like hunter-gatherers); an aging revolution (to use the time assigned to us by the genetic clock in a healthier way); and above all, the education revolution.

There is widespread agreement that Neolithic man was identical to us from a cerebral point of view. In a certain sense, every child born today is a Neolithic child, but through education, he or she rapidly becomes a modern adult.

Similarly, adequate education could in a relatively short time eliminate the profound differences that exist between human groups. In fact, the most important disparity is in knowledge, that is, level of education, which in turn determines many of the other differences between human beings and between nations.

Only a (permanent) educational revolution of this kind will allow humanity to develop in a balanced manner, so that culture and technology can be mutually enriching.

INTERVENTION IN HUMAN DEVELOPMENT

Other "revolutions" may pose more complex problems: for example, a longer life expectancy or a change in certain human features.

Today, the general trend among bioethical committees is to oppose any kind of genetic manipulation that alters the lineage of the human race in a hereditary way (i.e., that directly or indirectly modifies the sperm or the egg and, therefore, the species).

There has been cautious opening into intervention on somatic cells (that is, cells of the individual, which do not affect offspring); in fact, treatment for very serious diseases could profit from such techniques in the not too distant future. But given the explosive development of knowledge and genetic engineering techniques, how will bioethical committees consider things in the year 6000, 73,000 or 284,000 (that is, in less than half an hour on our calendar?)

As soon as we adopt a longer-term view, problems suddenly take on a different perspective and it becomes clear that our values could no longer be valid in totally different contexts (just as our current values would prevent us from abandoning our elderly or disabled in a forest or on the ice pack, something which was accepted behavior among nomadic hunting groups, including Indians and Eskimos).

If human beings succeed in changing their "biological clock"—as seems plausible—making it possible for them to enjoy good health for a few more decades or even centuries, bioethical committees will be faced with a completely new predicament: keeping human beings from living too long. A society of centenarians would be a real problem because life (and culture) require constant turnover and renewal, unless brains develop into something quite different from what they are today.

Natural selection has been altered

We are instinctively reluctant to accept that human beings could change in the future, that they could become different (perhaps even very different). But if we consider our past, we must recognize that we ourselves are quite different from our ancestors of 200,000, one million, or 10 million years ago (at that time, our ancestors were not yet bipedal apes). If evolution has produced such enormous changes in a single "day" (December 31st), what will have happened by next January 1st (i.e., in the next 10 million years)? Paradoxically, to remain as we are, we will have to intervene to stabilize evolution.

Actually, we are already acting upon evolution in a rather noticeable way: we have changed the rules of natural selection. Just look at the way we have changed plants through crossbreeding and selection. We have done the same to animals: not only domestic animals such as the dog (by producing such diverse breeds as the great Dane, the dachshund, the Saint Bernard, and the Pekinese from the same original species), but also farm animals such as the cow (some cows bred for milk production and life in automated barns would no longer be able to live in the wild).

Natural selection of human beings has also been altered. Many of the infants that would have been eliminated by nature in the past—even those with serious diseases and genetic defects—survive today and some of them become reproductive adults.

Natural selection no longer operates on the brain either. In the past, the human groups with more highly developed brains gradually became dominant, replacing other groups (and allowing for the slow emergence of *Homo sapiens sapiens*); today, selection for intelligence has ended. No one dies (or fails to reproduce) because they are not sufficiently intelligent. Therefore, the evolution toward more developed brains should not be expected to continue naturally, as in the past.

While human society is now being expressed through the values of solidarity, eliminating the brutal selection of the past—but also the positive values of selection—a way has been found to reestablish the former balance: human beings are entering into the "control room" of their own biology. It is to be expected that they will start to press those buttons, sooner or later, endowing those who have been "saved" with more health or, perhaps, starting a new kind of evolution based on the performance of the nervous system.

REMOTE-CONTROLLED EVOLUTION

What seems certain is that if human beings continue to sit in the planet's "control room," evolution will not continue along the same path it has taken for the last 4 billion years. This will be true for plants, animals, and human beings.

Future human societies may choose a "stationary," conservative model, maintaining nature and the human species the way it is (but that would require intervention in evolutionary mechanisms, which would tend to proceed in other directions in the long run). Furthermore, they could decide to recognize the importance of nature, rehabilitating ecosystems that have been destroyed by previous generations: forests, woods, lakes, perhaps even swamps, and their animals (the problem of global overpopulation could be solved in a relatively short time if a few generations decided to have only one child per couple).

But it is unlikely that things will remain as they are in the long run; it is far more probable that evolution will take quite a different tack.

ASSEMBLY CONTINUES

To try to imagine what evolution could be like in such a distant future, it may be best to look at the past.

As we have repeatedly stated, what has taken place over the last 4 billion years is a gradual "assembly" from the simple to the more complex. In order to deal with problems of adaptation to the environment, certain evolutionary lines became more and more complex and sophisticated (while others remained simpler).

It all started with atoms. The next step was molecules, then macromolecules like DNA and proteins contained in vesicles. This led to bacteria-like cells, then nucleate cells (deriving, at least in part, from an association of bacteria), colonies of cells, pluricellular organisms, and groups of pluricellular organisms.

What was the central element of this evolution? It is hard to say. It has been suggested that a hen is no more than a strategy adopted by an egg to produce another egg. Similarly, one could say that a cell is no more than a strategy adopted by the first bacteria for survival and reproduction. Or that primitive bacteria are no more than a strategy adopted by genes to reproduce other genes. The same holds for genes and DNA.

The answer should, perhaps, be sought in a totally different direction. The central element of assembly is assembly itself, that is, the structure. As the level of structural organization increases, the constituent parts lose their ability for individual designs as they are inserted into an integrated system which is moved by different mechanisms. This structure is something quite different from the sum of its single parts. As in language, the letters, the syllables, the words take on different roles and meanings in relation to the structure in which they are used.

If "assembly" (i.e., greater complexity) is the basic mechanism that has characterized 4 billion years of evolution, is it unreasonable to suppose that it may continue into the next billion? But how? With what pieces? We can only attempt to imagine it.

NEW PIECES ARE ADDED

Human beings have already added some very innovative "pieces" to biological evolutionary assembly. For a start, they have created something that has outclassed the human mind: the product of all human minds—culture. Human beings have created a cerebral "superstructure" that is far superior to any single human brain. The importance of the transition from a single mind to a pluricerebral intellect may be considered equivalent to that of the passage from unicellular to pluricellular organisms.

This network of brains or nervous systems led to a new and more complex structure that made possible the proliferation of knowledge, a turning point in evolution.

And this new kind of assembly has led to another turning point: the association of machines with the human nervous system. This is the first time that living beings have amplified their natural performances artificially. By joining with machines, human beings can outpower any animal, fly higher than any bird, dive to the depths of the seas, see the rings around Saturn, or communicate instantaneously with someone on the other side of the earth.

Like artificial organs, machines allow us to see protozoa and bacteria. They can make particles collide. They can also carry out calculations that no brain (or even group of brains) would ever be able to accomplish. A living system has created artificial "extensions" and machines that are beginning to imitate human brain processes and movements. Some of their calculation structures are even based on certain biological principles (e.g., biochips).

This may be the kind of assembly that will develop in the near future, involving methods, instruments, and levels of integration that we cannot even imagine today. In the remote future, other kinds of assembly using currently inconceivable "pieces" will follow.

Even science fiction is still in the Paleolithic as regards the new lines of development for the next billion years.

And what would happen if human beings disappeared before all this happens? In that case, it is possible (some scientists think even highly probable) that some other living form would retrace the route taken by human beings tens of millions of years earlier toward the development of intelligence, knowledge, culture, and technology.

What are the chances of this happening? No one can say. It could be pure fantasy; then again, reality could surpass our imagination. If someone landing on the earth 4 billion years ago had asked what the chances were of the primitive bacterial forms of that time developing into philosophers, painters, party leaders, or scientists, they would probably have received the same answer.

EXPANSION INTO THE COSMOS

Another fate also awaits the human species (or its descendants), and that is expansion into the cosmos.

We will not go into the conquest of the solar system or space colonies with millions of inhabitants (already technically conceivable today). Nor the energy or propulsion systems that could allow for interplanetary travel (such trips would certainly not be carried out by crews, but by immense colonies, that is, by small, self-propelled artificial planets). There is one aspect of these subjects, however, which is closely related to the theme of this book, and that is the possibility of the evolution that took place on Earth having also taken place on some other planet in the universe.

Numerous books are available on this topic and it is beyond our scope here to delve into the many hypotheses and debates raging in the scientific community. The most generally accepted opinion, however, is that the possibility does exist and that it may have led, as on Earth, to the birth of a technologically advanced civilization.

Years ago, NASA launched a program called Search for Extra-Terrestrial Intelligence (SETI), which used radio telescopes to listen into space and intercept possible radio communications coming from other planets. The program was very limited, but one of its indirect objectives was to make people aware of the possibility of extraterrestrial life. Of course, no one knows when and where (and if) such extraterrestrial civilizations may have arisen. But if such an event did take place (one or more times), we can at least imagine how.

The laws of physics and chemistry seem to hold throughout the universe; therefore, similar situations produce similar results. But of course, the question is: are there planets similar to Earth in the cosmos?

As the light reflected by a planet is too weak to reach us, no planet outside the solar system has ever been observed directly. It should not be forgotten, however, that there are as many as 100 billion suns in our galaxy alone (and there are a billion galaxies in the cosmos). Therefore, there are certainly many other solar systems in the universe similar to ours.

PLANETS SUITED TO LIFE

Today we know that a planet must be the right distance from its sun to support life (neither too close, so as not to be roasted, as on Mercury, nor too far away, so as not to freeze, as on Mars). It must also be the right size: neither too small, or it will not have enough gravity to hold its atmosphere, nor too large, or it will become a gaseous sphere, like Jupiter and Saturn.

As the composite picture of a planet suited for life emerges in increasing detail, it becomes clear that such a planet would be very similar to the earth: one bombarded by vagrant asteroids and rock fragments carrying water and chemical compounds useful for molecular "assembly." In other words, a planet of this kind would act like the closed vessel used by Miller (see chapter 1) to reproduce reactions similar to those that took place on the primordial earth. Naturally, no one can say whether these conditions are enough to trigger the development of life; but then again, in the only experimental situation that we know of—the earth—it worked.

Given the almost infinite number of possible planets, there is a statistical probability that extraterrestrial life does exist. And once the spark of life has been set off, it may lead to increasingly complex and perhaps even intelligent forms, if enough time is available.

As for the question of whether there is life elsewhere now, all we can say is that our sun is neither old nor young; it is middle-aged. Our galaxy has both suns that are older and others that are younger. Therefore, life could have evolved much (billions of years) earlier on certain planets and perhaps already have disappeared, or it could still be in a primordial stage on others. In any case, it is stimulating to think that we may not be alone in the cosmos and that some other planet has followed or is following the same path as ours.

As for the forms of life that could develop elsewhere, the mechanism of evolutionary convergence could come into play here: similar solutions to similar problems. Therefore, certain living forms on other planets could appear to be not too different from some forms on Earth. Actually, the variety of living forms and of environments and climates on Earth is so vast that the "repertoire" of possible adaptive responses has largely been tried out in one way or another. In addition, experience on Earth has proved that there are also many "cultural convergences" between distant populations in response to common requirements (writing, mathematics, navigation, agriculture, astronomy, and so on).

Will human beings ever be able to meet or even get in touch with such possible companions of extraterrestrial evolution? A NASA study has shown that the possibility is very small, but not nil. Radio signals and interplanetary probes, although in a very rudimentary way, are being sent into space. What will we be able to do or intercept in ten thousand or a million years' time? Some believe that we will be able to fill our galaxy with self-replicating probes; others predict even more extraordinary events, such as a remodeling of the solar system, including the demolition and reconstruction of certain planets.

It's a pity that we cannot imagine the future and the possible repercussions of a cosmic evolution of this kind. Indeed, until now we have spoken of projections of "only" hundreds of thousands or at the most a million years into the future. That is no more than a few days past January 1st on our calendar. What will happen in March, June, September, or December of the next evolutionary year (i.e in 1, 2, 3, or 4 billion years)? Or beyond?

THE EARTH IS KILLED BY THE SUN

Current knowledge allows us to make one rather precise prediction in this time frame: in approximately 5 billion years, the earth will be burned by the sun. In fact, once it has burned out its fuel (hydrogen), the sun will enter into a new phase, that of a red giant, that is, a dying star, which will rain down enormous quantities of heat and light on the earth before becoming a white dwarf.

At that point, the oceans will boil and evaporate, destroying the atmosphere. It will become so hot that rocks will melt, devastating the planet. Not only will all forms of life disappear, but all traces of human culture on Earth—architecture, works of art, archives, cities, archeological remains—will be wiped out (but here, too, we are incapable of imagining what the response of intelligent life might be).

It is very probable that humans will no longer be present on the planet. Long before this time, they (or their unrecognizable descendants) may have moved elsewhere, assuming of course that they have not been extinct for billions of years. Therefore, cosmic developments will still condition the evolution of the earth very strongly. Born with the formation of the solar system, the earth will be destroyed by the same star which made its development and the evolution of life upon its surface possible.

And then what will happen?

Our knowledge today suggests that all suns, that is, all stars in all galaxies, will undergo the same fate, although at different times. In its final stages, therefore, the ever expanding universe will look like the sky after a fireworks show, with its exploding showers of stars slowly giving way to total darkness. The stars will go out, one by one, like theater lights, and the protons themselves (i.e., matter) may decay and disintegrate. There will be no place to go, in spite of the most extraordinary technology, as it will be very difficult to overcome the Second Law of Thermodynamics, which calls for a final zeroing of potential energy. Unless . . .

Unless there are enough neutrinos (or other more "exotic" particles) in the universe for their "hidden mass" to make it collapse onto itself (the way a handful of gravel thrown into the air falls back to the ground because of the pull of gravity).

In that case, no one can imagine what would happen. Some scientists believe that if the universe collapsed onto itself it would return to a condition of high density and high temperature, perhaps resulting in another Big Bang followed by expansion. This is only a hypothesis, but it is obviously much more attractive than the first. Like the proverbial phoenix, the universe could rise again from its ashes and start out on a new adventure.

If that were the case, a new sky would be populated with new galaxies, new suns, and new planets. And it would be nice to think that, one day, on one of these planets, two hypothetical travelers could land to tell the story of the birth of life on their planet and its slow but extraordinary evolution.

But there is also a third hypothesis: that the mass of the universe is exactly equal to the "critical mass," setting it halfway between the two conditions described above (modern cosmological data seems to point in this direction). The universe, in this case, would continue to expand at an increasingly slower rate, tending to come to a stop, but without ever collapsing. The scenario would be one of a long, ever darker night in the midst of an expanse of dead or disintegrating galaxies that drift off into space. Their light, weak and remote, would continue to travel in the cosmos for a very long time after their death.

THE RETURN TO EARTH

Our trip has come to an end.

From the porthole of our spaceship, we can see the earth approaching. The crew is busy preparing for the maneuver: the captain is verifying the angle of inclination for reentry into the atmosphere with the control center. On the one side is the sun, radiant against a black, star-speckled sky, a profound universe which still conceals many of its secrets and which may be the stage of a new phase of human evolution. On the other is our planet, with its blues and whites and its precious cargo of life hidden behind the atmosphere's veils. Also shrouded are the oceans, where the process that led to the emergence of plants, animals, and human beings over a period of 4 billion years began. That is the extraordinary story that we have tried to recreate and tell.

Whatever the current and future problems of the earth, the transition phase that we are now going through is only a microsecond on the evolutionary clock. When set into that kind of time frame, all presumption appears ridiculous.

It may be comforting to think that many things, places, and people that irritate us today will inexorably be wiped away, like evolutionary refuse. But it is also disconcerting to see how we risk jeopardizing the quality of development and triggering crises that could be avoided through a more intelligent and creative management of technology as well as through a greater educational effort, putting us abreast with and possibly setting us ahead of the pace of change. After all, this is the time in which we have to live.